Quality Assessment of Milk and Milk Products

NIPA® GENX ELECTRONIC RESOURCES & SOLUTIONS P. LTD.
New Delhi-110 034

Quality Assessment of Milk and Milk Products

D.K. Thompkinson
Dairy Technology Divison
National Dairy Research Institute
Kamal (Haryana) India

NIPA® GENX ELECTRONIC RESOURCES & SOLUTIONS P. LTD.
New Delhi-110 034

nipa®

NIPA® GENX ELECTRONIC RESOURCES & SOLUTIONS P. LTD.

101,103, Vikas Surya Plaza, CU Block
L.S.C. Market, Pitam Pura, New Delhi-110 034
Ph : +91 11 27341616, 27341717, 27341718
E-mail: newindiapublishingagency@gmail.com
web: www.nipabooks.com

For customer assistance, please contact
Phone: + 91-11-27 34 17 17
Fax: + 91-11- 27 34 16 16
E-Mail: feedbacks@nipabooks.com

ISBN: 978-81-19254-33-0

NIPA® also publishes books in a variety of electronic formats. Some content that appears in print may not be available in electronic books, and vice versa.

Composed and Designed by NIPA®.

Contents

प्रोफेसर (डा.) ए.के. श्रीवास्तव
निदेशक
Prof. (Dr.) A.K. Srivastava
Director

राष्ट्रीय डेरी अनुसंधान संस्थान
NATIONAL DAIRY RESEARCH INSTITUTE
(मान्य विश्वविद्यालय)
(Deemed University)
(भारतीय कृषि अनुसंधान परिषद)
(Indian Council of Agricultural Research)
करनाल–132 001, (हरियाणा) भारत
Karnal-132 001, (Haryana) India

Foreword

India has emerged as the largest milk producer in the World with total of 114 million tones. Dairying as part of integral -agriculture system had always plays an important role in agriculture economy. It is a major source of income to the small, land less and marginal farmers who consider milk as a cash crop. Majority of milk producing sheds are located in the rural area far away from me point of processing. Such milk seldom reaches timely to the scarcely existing chilling centers. The quality of milk being highly cashable, may undergo deterioration very fast, if not processed thin permissible time limits.

The universal rule of producing safe and quality milk products 'to have good quality of raw material-MILK. As the reach of organized dairies to farmer door step for milk collection is limited, most of the raw milk from rural areas is picked up by the contractors who is neither aware nor give any importance to the quality of milk. Some contractors even follow unscrupulous practices of adulterating the milk which ultimately ends up at the processing units.

In the present era of globalization with boarder-less trade World over. the Indian dairy industry has to compete to produce safe and quality products at par with other developed countries. In this context quality parameters and quality management systems have become more imperative. Even though India has become largest milk producer, it still lab behind considerably on the quality front. Monitoring of quality at every stage of production, procurement, processing and marketing, therefore, needs to be looked into carefully and thoroughly.

The present publication is an important step forward in delineating possible techniques in ascertaining quality, at every step from production to processing of milk and milk products, to ensure consumer safety. It also touches various aspects of different food safety laws and acts *vis-a-vis* international standards for implementation of quality management system. It is a good attempt in consolidating various techniques/ methodologies as a quick reference for the quality assurance personals associated with dairy industry as well as to the new technocrats, being trained at various dairy science colleges, as a human resource for our dairy industry.

Dr. A.K. Srivastava
Director, NDRI, Kamal

Preface

Dairy industry is an important segment of food industry in the country. The food business environment has under gone a sea change after the trade reforms initiated by WTO. Production of dairy products using good manufacturing practices is not sufficient until the product is made safe for the intended consumer. Food safety has been recognized as *de rigueur* for survival under competitive global market. The WTO agreement has made the World smaller for trading. Imported dairy and food products are flooding the Indian market resulting in stiffer competition for the local industry. India also needs to enter into World market with its own brands of dairy products and a beginning has been made by the Indian dairy industry in this direction. Although, India has achieved the status of largest milk producer, on quality front it lags behind considerably. If India has to make big stride in the export of dairy products to the other developed countries it becomes essential to harmonise with the existing standard procedures in the World trade.

In the borderless world trade it has become imperative to signify the quality capability of a company by quality systems. In this direction it has become very essential to obtain necessary international recognition to exhibit our commitment to quality. To develop and manufacture globally competitive food product it becomes imperative to establish testing laboratories and centre for excellence for R&D. In order to manage production and quality of processed food, it is equally important to generate technically professional manpower through support to training institutions.

The present monograph is an attempt to compile various chemical. Nutritional and micro-biological analytical techniques related to handling milk and milk products. The monograph has been divided into three basic sections dealing in methodologies applicable to raw milk reception to milk product analysis to air, water detergents and dairy effluent. Section "A" deals with various plateform and laboratory tests applied to raw and processed milk quality assessment, in terms of rapid tests, detection of adulterants and preservatives, sampling techniques *etc.* Section "B" gives the detailed analytical techniques as applied for determination of various quality parameters/ attributes of different dairy products being manufactured by Indian Dairy Industry. In section "C" various other tests applicable to dairy industry for maintaining quality including nutritional quality and hygiene, test for dairy detergents and sanitizers, air and water quality for a dairy plant and determination of malamine, aflatoxins, minerals and vitamins (fat & water soluble).

Apart from coverage of various analytical techniques for maintaining quality of different dairy products, the text also contains various food regulatory acts, food safety act 2006, hygiene code, ISO 9000 series, ISO 14000 and ISO 22000 quality management systems, and BIS standards for different milk types and MRL values for pesticides, heavy metals and antibiotic substances and list of approved analytical labs are also provided for ready reference.

It is expected that such a compilation of all possible chemical, nutritional and micro-biological techniques as applied in management of product quality will be useful both for students as well as professionals under training or in respective jobs at various dairy plants across the country.

Author

December, 2011

Acknowledgement

The author is indebted to the editors and publishers of variety of books and journals from which meaningful information has been gathered and compiled in this presentation. The author also gratefully acknowledge the contribution of friends and colleagues viz. Dr. Patel, Dr. Jessa Ram, Mrs Savitri Jhamb, Mrs. Prabha Sharma, Dr. S.K. Anand, Dr. N. Goel, Dr. R. Sharma, Dr. Patil, Mr. Manoj, Dr. Singla, Mr. K.L.Arora and Mr.Garg, at the department of dairy technology, dairy microbiology and dairy chemistry and Export inspection agency, New Delhi who have from time to time shared their expertise in different aspects related to various techniques / methodologies and standards *etc.,* during preparation and presentation of this book into existing form.

The author would like to express his deep sense of gratitude to Dr. B.N. Mathur, Ex-Director, National Dairy Research Institute, Kamal for his valuable suggestions during compilation of subject matters and their presentation. The author is also indebted to Dr. Sushil Kumar, Ex-Director, N.D.R.I., Karnal and Dr. A.K.Srivastava, present Director, N.D.R.I., Kamal for their encouragement and permission to write this book.

The references from various books, journals, compilations, notes and presentations have been freely consulted, for which the author is gratefully acknowledge the publishers and editors. Thanks are also due to my family members, well-wishers, friends and relatives for their continued support in this endeavor.

Last but not the least the author extends his heartfelt thanks to the publishers of this book, since without their kind co-operation this book would not have seen the light of the day.

D.K. Thompkinson

December 2011

Introduction

Quality may be defined as degree of excellence. It is generally understood in terms of excellent product, which fulfills or exceeds customers expectation. Quality is a degree to which a set of inherent characteristics fulfills requirement (ISO 9000-22000) which may be quantitative or qualitative. The word quality has multiple meaning but those, which dominate the use of this word, are freedom from deficiency and product features that meet customer needs and satisfaction.

Quality has different dimensions and may be measured using various parameters like- Aesthetics, features, conformance, response, reliability and reputation. Quality has become an important aspect and essential part of any organization to sustain, progress and upgrade its product quality. Today the customer has become extremely concern about the quality and safety of products rather than the price.

In the present era, quality and stringent standards for milk and milk products posses greater challenge to the Indian Dairy Industry. In an open market policy it is increasingly difficult to compete with global players and survive unless the quality of high standards is maintained consistently. Globalisation of food trade as a result of liberalized trade policies has increased the risk of cross-border, infections. Food safety and quality assurance issues are therefore, n the highest priority list of government agencies, producers and manufacturers alike.

Chapter–1

Quality Assessment of Raw and Processed Milk

Milk is a product of biological origin and is a suitable vehicle to accommodate any additive without apparent changes in its look. The quality of milk and its products depends upon quality of raw milk used in their manufacture, processing and handling conditions. It is therefore, necessary to check quality of raw milk to assess its suitability for processing, through various quick tests called "Platform Tests"

Raw milk quality assessment

(A) At collection center: - Milk from various suppliers is generally collected at dairy factory or privately owned collection centers located in marked area of milk sheds. The quantum of supply may vary from 2 – 100 Liters. As each supplier has to be paid in relation to quantity and quality of his produce, it is important to draw samples and ascertain its quality by performing quick tests for the purpose of acceptance or rejection.

Sampling procedures - The sampling procedure differs according to the nature of material and the purpose for which the sample is required. The sample collected for chemical analysis should be representative of the entire batch. Thorough mixing with proper instrument is essential to ensure representative sample. Samples for chemical analysis are to be delivered the same day for testing. If un-avoided, add small amount of preservative (Formaldehyde) to prevent deterioration and bear label to the effect.

Sample from single container – The procedure involved is thorough mixing of contents. If it is small container the contents should be poured from one vessel to another with swirling. In case of can the contents should be mixed with the help of plunger which should be moved up and down and sideways for about 10-15 times. The sample is then drawn with the help of dipper from different levels and collected in clean and dry sample bottles.

Rapid Platform Tests

(i) Visual and Organoleptic test

Aim :- To reject the milk having objectionable flavour and colour.

Procedure

1. Remove lid from the can / tanker and observe for any extraneous matter present.
2. Sniff the milk for any objectionable flavour that is un-natural of milk.
3. Put small quantity (20 – 30 ml) on tongue and roll into mouth cavity for any off taste that is un-natural of milk.
4. In case of doubt subject the sample to other tests.
5. Based on organoleptic evaluation accept or reject the milk.

(ii) Sediment Test

Aim :- To asses the cleanliness of milk received.

Preparation of standard disc

1. Weigh 53g and 10g of cow/buffalo dung (passed through

425 and 850 micron sieve respectively); 27g garden soil (passed through 425 micron sieve); 8g and 2g charcoal (passed through 425 and 850 micron sieve respectively).

2. Dry these materials in an oven at 100° C and make a uniform mixture.
3. Weigh 0.1g of the mixture and transfer into 1000 ml volumetric flask.
4. Using 50% sugar solution, wash all particles into the flask.
5. Makeup the volume with sugar solution and mix the contents.
6. Vigorously shake the contents every 5 minutes for the next 50-60 minutes to thoroughly wet the particles.
7. The mixture is ready for disc making after the particles are evenly suspended.
8. Ten milliliter of this solution will represent 1mg of sediment.
9. Place 150ml or more of filtered skim milk in sediment tester.
10. Add varying amounts of sediment suspension and force through the sediment tester.
11. Take small amount of skim milk and pass through sediment tester to obtain even distribution of sediment on the disc.
12. Prepare five such discs showing 0.0; 0.2; 0.5; 1.0 and 2.0mg sediment.
13. Remove the discs, mount on stiff paper and allow to dry.
14. Make permanent by spraying strong disinfectant.
15. Note quantity of dried material below each disc for purpose of comparison.

Procedure

1. Assemble the off bottom sediment tester, for cans and tankers, by inserting sediment disc in the space provided.

2. Dip the tester to the bottom of the can without disturbing the milk.
3. Collect milk slowly from different parts of can bottom by pulling the plunger upwards.
4. Lift the sediment tester and press the plunger down and empty the tester.
5. Dismantle the sediment tester and remove sediment disc carefully.
6. Compare with standard sediment disc or weigh the disc.
7. Grade milk in accordance to ISI standard.
8. In case of mixed sample sediment tester pass 500 ml milk through the tester.

BIS Standard for Sediment

Amount of sediment	Grade
0.0 mg	Excellent
0.2 mg	Good
0.5 mg	Fair
1.0 mg	Poor
2.0 mg	Very Poor

(iii) Colt-on boiling Test (COB)

Aim :- To asses the suitability of milk for pasteurization

Procedure

1. Transfer 5.0 ml of milk in a 15 ml test tube using graduated pipette.
2. Place the tube in boiling water bath for five minutes.
3. Alternatively, hold the tube over a flame and allow the contents to boil.
4. Observe formation of clots or flakes on the test tube wall.

5. Formation of flakes or clots indicates positive test.

Note :- Presence of clots or flakes show that the milk has high developed acidity and is not suitable for pasteurization.

(iv) Alcohol Test

Aim :- To asses the suitability of milk for high heat processing

Procedure

1. Transfer 5.0 ml of milk in a test tube using graduated pipette.
2. Add equal amount of ethyl alcohol (75% for cow milk and 65% for buffalo milk).
3. Close the mouth of the test tube with thumb and mix the contents well by inverting the test tube several times.
4. Observe for formation of any flakes on the wall of the test tube.
5. Presence of flakes indicate positive test.

Note :- Flake formation shows that the milk is abnormal either due to high acidity or disturbed salt balance. Such milk is not suitable for high heat treatment.

(v) Alcohol –Alizarin Test

Aim :- To asses suitability of milk for high heat treatment and to have idea about milk acidity without acidity test.

Procedure

1. Transfer 5.0 ml of milk in a test tube using graduated pipette.
2. Add equal amount of alcohol-alizarin solution (0.2%).
3. Mix the contents well by inverting the test tube several times.
4. Observe for formation of flakes and colour of the contents.
5. Match your results with following table

Range of colour	Presence of flakes	Approximate Acidity (%)
Red to Yellow	—	0.18 – 0.36
Brown red	Nil / no	0.16
Reddish brown	small	0.20
Yellowish brown	small	0.24
Brownish yellow	large	0.28
Yellow	large	0.36
Violet		Alkaline

(vi) Neutralizer test

Aim :- To detect presence of added Carbonate or Bi-carbonate in milk.

Procedure

1. Transfer 5.0 ml of milk in a test tube.
2. Add equal amount of 75% ethyl alcohol.
3. Add few drops of 1% aqueous solution of Rosalic acid.
4. Mix well and observe the colour.
5. Presence of rose red OR pink colour indicates addition of neutralizer in milk.

Note :- Rose red colour will appear upon addition of - NaOH; KOH; $Ca(OH)_2$ *While Pink colour will appear upon addition of* $NaHCO_3$*;* KCO_3*;* $CaCo_3$

Confirmatory test

1. Transfer 5.0 ml of milk to 150 ml conical flask.
2. Add two drops of conc. HCl and connect to a glass tube bend at 90-degree angle.
3. Pipette 2 ml of 0.001N NaOH in a test tube and add few drops of phenolphthalein indicator.
4. Dip the glass tube in solution of NaOH.

5. Heat the contents of conical flask to boiling for 3 minutes and pass the evolved CO_2 through NaOH solution in tube.
6. If the red colour turns colourless, the test confirms presence of carbonates / bicarbonates.

Detection of preservatives

Milk and milk products are required to be free from any kind of additives from public health point of view. The law therefore, does not allow addition of any kind of preservatives in milk. It can only be added in samples for purpose of testing. However, some producers add these additives in order to protect their produce from deterioration due to delay in transportation of milk at the dairy plant. The practice is especially prevalent in tropical countries like India. The common preservatives added are boric acid, benzoic acid, formaldehyde/ formalin and some times it comes from chlorine sterilizers, used for sterilization of milk handling equipments, in the form of hypochlorite residue.

(a) Test for boric acid

(i) Aim :- To determine presence of **boric acid and borates** in milk.

Boric acid and its salts give characteristic red colour with turmeric paper.

Turmeric paper test

Procedure

1. Transfer 5.0 ml of milk in a test tube.
2. Add 1.0 ml of concentrated hydrochloric acid and mix well.
3. Dip a strip of turmeric paper in the acidified milk.
4. Dry the paper and note the change in colour.
5. Red colour indicates presence of boric acid.
6. Add a drop of ammonium hydroxide solution.

7. Change from red to dark green colour confirms presence of boric acid.

Glycerol Test

Procedure

1. Take 10.0 ml of milk in a test tube.
2. Add few drops of 1% phenolphthalein indicator and mix well.
3. Add drop wise 0.1N NaOH till faint pink colour is obtained.
4. Divide neutralized sample into two tubes.
5. To tube one add equal volume of 50% glycerol solution and to another add water.
6. Compare the colour of two tubes.
7. If the tube containing glycerol is lighter in colour, it indicates presence of boric acid.

(b) Test for Formaldehyde

(ii) Aim :- To detect presence of **formalin or formaldehyde** in milk.

Formaldehyde gives characteristic violet colour with ferric salts and other oxidizing agents. There are mainly two tests viz. Hehner and Lech test for such determinations.

(a) Hehner Test

1. Take 10.0 ml of milk in a test tube.
2. Add 0.5 ml of 10% ferric chloride solution.
3. Add carefully, 5.0 ml of concentrated sulphuric acid down the side of test tube in such a way that it forms a separate layer at the bottom without mixing with milk.
4. Observe the colour of ring formed at the junction of two liquids.

5. Presence of violet coloured ring indicates presence of formaldehyde.

b) Lech Test

1. Take 5.0 ml of milk in a test tube.
2. Add equal volume of concentrated hydrochloric acid containing 1.0ml of 10% ferric chloride solution (500 ml HCl + 1 ml FeCl3).
3. Heat the contents over a flame for 5 minutes.
4. Rotate or shake the tube to break the curd.
5. Observe the colour.
6. Development of violet colour indicates presence of formaldehyde.

(c) Test for H_2O_2

(iii) Aim :- Detection of added **hydrogen peroxide** in milk.

Paraphenyle-diamine-hydrochloride gives intense blue colour with hydrogen peroxide.

Procedure-1

1. Transfer 10.0 ml of milk in a test tube.
2. Add 2 drops of 2% freshly prepared aqueous solution of paraphenyle- diamine- hydrochloride.
3. Observe the colour.
4. Development of intense blue colour indicates presence of hydrogen peroxide.

Procedure-2

1. Transfer 1.0 ml of milk in a test tube.
2. Add 1.0 ml of potassium iodide – starch reagent (20% KI and 1% starch solution, mixed in equal volume).

3. Observe the colour.
4. Appearance of blue colour indicates presence of hydrogen peroxide.

(d) Test for Benzoic Acid

Aim :- Detection of added **benzoic acid** in milk.

Procedure-1

1. Transfer 10.0 ml of milk in a test tube.
2. Add 5ml of dilute hydrochloric acid (1:3) and mix well.
3. Filter and extract the filtrate with 50-100ml ethyl ether.
4. Wash ether layer with two 5ml portions of water.
5. Evaporate ether in a porcelain dish on a water bath to dryness.
6. Dissolve residue in hot water. Add 2 drops of 10% NaOH solution and evaporate to dryness.
7. Add 5 drops of concentrated sulphuric acid and a small crystal of potassium nitrate.
8. Heat for 20 minutes in boiling water bath. Cool and add 1ml water, make distinctly ammoniacal and further heat for 20 minutes.
9. Cool and add a drop of freshly prepared ammonium sulphide without allowing layers to mix.
10. A red-brown ring indicates presence of benzoic acid.

Procedure -2

1. Transfer 10.0 ml of milk in a test tube.
2. Add 5ml of dilute hydrochloric acid (1:3) and mix well.
3. Filter and extract the filtrate with 50-100ml ethyl ether.
4. Wash ether layer with two 5ml portions of water.

5. Evaporate ether in a porcelain dish on a water bath to dryness.
6. Dissolve residue in hot water and few drops of 0.5% neutral ferric chloride solution.
7. A salmon coloured precipitate indicates presence of benzoic acid .

(e) Test for Salicylic acid

Aim :- Detection of added **salicylic acid** in milk.

Procedure

1. Transfer 10.0 ml of milk in a test tube.
2. Add 5ml of dilute hydrochloric acid (1:3) and mix well.
3. Filter and extract the filtrate with 50-100 ml ethyl ether.
4. Wash ether layer with two 5ml portions of water.
5. Evaporate ether in a porcelain dish on a water bath to dryness.
6. Dissolve residue in hot water. Add 2 drops of 10% NaOH solution and evaporate to dryness.
7. Add 1-2 drops of 0.5% ferric chloride solution.
8. A violet colour indicates presence of salicylic acid.

(f) Test for Hypochlorite

Aim :- Detection of **hypochlorite** in milk.

Procedure

1. Transfer 3.0 ml of milk in a test tube.
2. In another tube take 3.0ml of 0.025% stannous chloride solution.
3. Place the tubes in freezing mixture of ice and salt for 3-4 minutes.

4. Take the mixture in centrifuge tube and centrifuge at 2500 rpm for 3-5 minutes.
5. Examine the tube for development of yellow-green colour.
6. Alternatively, examine the tube under UV light of mercury vapor lamp fitted with woods filter.
7. Yellow fluorescence confirms presence of hypochlorite.

Detection of adulterants

Milk is a fluid where any additive can be added without bringing any changes in its physical appearance. These additives are generally added to defy lactometer tests, generally applied to check any deviation in specific gravity or solid-not-fat content of milk sample. A number of additives are used in maintaining the lactometer reading up to the mark to avoid suspension towards adulteration. However, these are not allowed by law and there detection is therefore important.

(a) Presence of added cane sugar

Resorcinol produces red colour with sucrose in acidic medium.

Procedure

1. Take 10.0 ml of milk in a test tube.
2. Add 1.0 ml of concentrated hydrochloric acid and mix well.
3. Add 0.1 gm of resorcinol powder and mix thoroughly.
4. Place the tube in a boiling water bath for 5 minutes.
5. Observe the colour of the contents.
6. Development of red colour indicates presence of cane sugar in milk.

(b) Presence of starch and other cereal flours

Iodine solution gives intense blue colour with starches

Procedure

1. Take 3.0 ml of well mixed milk in a test tube.
2. Boil the contents over Bunsen burner.
3. Cool and add one drop of 1% iodine solution
4. Mix well and observe the colour.
5. Development of intense blue or bluish violet colour indicates presence of added starch.

(c) Presence of urea

Urea is added to milk in order to increase nitrogen content and thereby corresponding protein content. Addition of para-dimethylamino-benzaldehyde produces intense yellow colour. Alternatively, presence of phenol in alkaline sample produces bluish-green colour.

Procedure 1

1. Take 5.0 ml of well mixed sample in a test tube.
2. Add 5.0 ml of para-dimethylamino benzaldehyde solution (DMAB).
3. Mix well and observe the colour.
4. Development of intense yellow colour indicates presence of urea.

Procedure 2

1. Take 3.0 ml of well mixed sample in a test tube.
2. Add 3.0 ml of 24% TCA solution.
3. Mix well and filter the contents through Whatman No. 42 filter paper.
4. Collect 3ml of filtrate and add 3ml of DMAB solution.
5. Mix thoroughly and observe change in colour.

6. Development of distinct yellow colour indicates presence of urea.

(d) Presence of glucose

Glucose is easily available commercially as concentrated syrup and is cheaper. It is being used as an adulterant in milk and for the manufacture of synthetic milk.

Requirement: **-** Test tube, boiling water bath, Barfords reagent, sodium acetate buffer (pH 4.75), phosphomolybdic acid reagent.

Barford reagent – Dissolve 24 g of cupric acetate in 450 ml boiling distilled water and add 25 ml of 8.5% lactic acid to the hot solution. Cool and dilute to 500 ml. Filter off impurities.

Phosphomolybdic acid reagent – Take 35 g of ammonium molybdate and 5 g of sodium tungstate in a one volumetric flask, add 200 ml of 10% NaOH and 200 ml of distilled water. Boil vigorously for 20-40 minutes and cool. Add 125 ml of concentrated phosphoric acid and makeup the volume.

Sodium acetate buffer – Mix equal volumes of 1N sodium acetate and 1N acetic acid solution.

Procedure

1. Take 1 ml of sample in a test tube and add 1 ml of Barfords reagent.
2. Heat the mixture on boiling water bath for 3 minutes.
3. Cool for 3 minutes under ice cold water.
4. Add 1 ml of phosphomolybidic acid reagent.
5. Formation of deep blue colour indicates presence of glucose.

(e) Presence of maltodextrin

Maltodextrin is used as cheap source of adulterant for enhancing sniff content of milk. The test is based on the release of glucose, after enzymatic reaction, which can be detected with simple diastic strip being specific for glucose.

***Requirement:* -** Beaker, graduated pipette (20x2 ml) , water bath, stop watch, pH paper (range 3.5-6.0), lactic acid, diastic strip(Bayer diagnostic India Ltd.)

Enzyme solution – Dissolve 0.2 g of alfa gluco amylase enzyme in 100 ml of distilled in a volumetric flask. Stopper and store under refrigeration. Use before 15 days.

Lactic acid solution – Take 10 ml of concentrated lactic acid in a 100 ml volumetric flask and make up the volume with distilled water.

Testing strip – Diastic strip for glucose. Store below 30°C but do not store under refrigeration. Use before six months.

Procedure - 1

1. Take 20 ml of milk sample in 100 ml beaker.
2. Adjust pH to 4.0 – 4.5 using 0.8-1.5 ml of lactic acid solution.
3. Add 1 ml of enzyme solution and incubate at 62°C for 5 minutes.
4. Cool contents to room temperature.
5. Remove one diastic strip from container bottle by holding the plastic end of the strip.
6. Dip the strip in test solution for 30 seconds and remove excess liquid by a single jerk.
7. Compare change of colour, from green to brown, with colour chart on the testing strip bottle and record the result.
8. Change of colour intensity (+, ++, +++, ++++) as depicted on glucose testing strip bottle indicate presence of maltodextrin.
9. Repeat test if incubation time exceed 5 minutes.

Procedure – 2

1. Take 5ml of suspected milk sample in a test tube.
2. Add 2ml of 0.05N iodine solution.

3. Appearance of red-brown colour indicates presence of maltodextrin.

(f) Detection of pond water

Procedure

1. Take 5 ml of milk sample in a test tube.
2. Rinse the test tube with sample and decant.
3. Add 1-2 ml of 2% diphenylamine solution along the side of test tube.
4. Appearance of blue colour indicates presence of pond water.

(g) Determination of specific gravity

Procedure

1. Adjust the temperature of milk sample to measuring temperature prescribed for lactometer (15.5° C).
2. Mix the sample well avoiding incorporation of air or foam formation.
3. Pour sufficient milk into the glass or steel cylinder to allow free floating of lactometer.
4. Place the lactometer in the milk and allow it to float till it stops and assumes a constant level.
5. Record the lactometer reading and temperature of milk at the same time.
6. Take another reading by flapping the top of the lactometer stem and when it again assumes constant level.
7. Take average of the two readings. Get corrected value (CLR) from the standard table for corresponding temperature.
8. Calculate the specific gravity of milk using the given formulas.

   Calculation :- Specific gravity = $CLR / 1000 + 1$

Processed milk quality assessment

(B) At dairy plant - All the milk that is collected at the collection center is generally chilled in bulk coolers before it is being transferred to cans or rail / road tankers and transported to the main processing station (Dairy plant).

The whole milk received at the dairy factory is processed to obtain cream and various categories of liquid milk as per demand and marketing strategies of particular dairy plant. To ensure supply of safe milk to the customers, the milk type is subjected to different processes and the effectiveness of process like pasteurization and sterilization needs to be checked upon to meet the legal standards for different milk categories in accordance to the standards prescribed by the regulatory agencies.

a) Fat / solid-not-fat / Total solid / Protein / Lactose/ Ash / Acidity

The procedures for above tests are similar as in case of raw milk testing. Detailed test procedure are delineated in chapter 2.

b) Phosphatase test

Raw milk contains phosphatase enzyme that is destroyed by efficient pasteurization. When milk containing phosphatase is incubated with p-nitro-phenyl disodium orthophosphate it librates para-nitro-phenol, which gives yellow colour under alkaline conditions. The colour is the measure of phosphatase content of milk.

Requirements: - Lovibond comparator, standard disc-APWT, glass cells, water bath, pipette, volumetric flask, measuring cylinder, test tubes, sodium carbonate, sodium bicarbonate, p-nitro-phenyl disodium orthophosphate.

Buffer solution – Dissolve 3.5 g anhydrous sodium carbonate and 1.5 g sodium bicarbonate in distilled water and makeup the volume to 1.0 litre in a volumetric flask.

Buffer-substrate solution – Transfer 0.15 g of substrate (p-nitro-phenyl disodium orthophosphate) in 100 ml volumetric flask and makeup the volume with buffer solution. Stopper the flask and store in refrigerator protected from light. This is a colourless solution, which should be used before one-week time.

Procedure

1. Transfer 5.0 ml of buffer substrate solution to a test tube.
2. Stopper the tube and place in water bath at 37° C.
3. After 10 minutes add 1.0 ml of pasteurized milk sample to the tube, stopper and incubate at 37° C for two hour.
4. Prepare a blank using boiled milk of same type and incubate at 37° C for 30 minutes.
5. Place the blank and the test sample in the lovibond comparator stand and view through central aperture of the disc.
6. Place the comparator facing good source of daylight. Revolve the disc and compare the colour.
7. Record the disc reading.
8. Interpret result as per given table.

Disc reading	Interpretation
0 to trace	Properly Pasteurized
2 to 6	Doubtful
10 & above	Under pasteurized

c) Test for sterilization

(i) Turbidity test – It is an indication of proper sterilization process.

1. Weigh 4.0g of ammonium sulphate in a conical flask.
2. Measure 20 ml of sample and transfer to the flask.
3. Mix the contents thoroughly and allow to stand for 5 minutes.
4. Filter through Whatman No 12 filter paper.
5. Collect 5 ml of filtrate in a test tube and place in boiling water bath for 5 minutes.
6. Cool the contents and examine for any turbidity.

7. Milk with no sing of turbidity has been sterilized satisfactorily.

ii) Alcohol precipitation test - It is an indication for assessment of keeping quality of sterilized milk at room temperature.

1. Incubate a sample of sterilized milk at 37°C for 24 hours.
2. Shake well and transfer 2ml of sample in a test tube.
3. Add 2ml of 80% ethyl alcohol and mix the contents.
4. Observe formation of flakes / clot.
5. Clotting indicates that keeping quality of processed milk is less than 21 days at room temperature.

d) Homogenization efficiency

Homogenized milk is the milk that has been treated to breakup fat globules into such small particles that they do not cluster and rise. The fat emulsion is stabilized so that no cream layer is formed. The tests indicate the stability of fat emulsion in homogenized milk. Usually a pressure of 2000-2500 lbs is sufficient to render milk adequately homogenized.

Procedure

1. Take a sample of homogenized milk into 250 ml glass bottle.
2. Allow to stand undisturbed for 48 hours in a refrigerator.
3. Carefully decant 100 ml of the upper layer, mix well and test for fat content.
4. Mix remainder of the sample and test for fat content separately.
5. Calculate the percent difference between the two tests.
6. The difference should not be more than 10%.
7. A value of 7-8% indicates most efficient homogenization.

e) Creaming index

1. Take 50 ml each of homogenized milk samples in three different graduated centrifuge tubes.

2. Centrifuge the contents for 15 minutes at 1000 rpm.
3. Transfer 5 ml upper portion from each tube into a beaker, using separate pipettes, and determine fat content of the same (sample I).
4. Empty the three tubes into separate containers mix well and determine the fat content (sample II).

Calculation :- Creaming index = (A – B) x 100 / B

Where A = fat content of sample I

B = fat content of sample II

Creaming index	Homogenization quality
Up to 10	Excellent
11 - 20	Good
21 - 30	Fair
Over 30	bad

f) Microscopic Method – (For Fat Globule Size Estimation)

1. Take 2 ml of solution A (4% sodium citrate and 0.2% NaOH) in a test tube.
2. Add 6 ml of solution B (10% gelatin and 1% phenol) and mix well. Heat to 65° C in water bath.
3. Diluted milk sample with distilled water in the ratio of 1:10 and add 2ml of diluted milk to the above solution.
4. Continue heating for another 5 minutes with mixing of contents.
5. Cool in an ice bath with gentle mixing until gelatin starts thickening.
6. Take small amount of mixture from the test tube with the help of platinum loop on a microscopic glass slide and cover with cover slide.
7. Place the sample slide onto microscope stage and put the micrometer in the eyepiece (15X).

8. Place stage micrometer under oil immersion lens.
9. Look through the eye piece and compare the divisions of the two micrometers and check how many divisions of ocular is equal to how many of stage micrometer.
10. Calculate the value of one ocular division.
11. (100 divisions = 1 mm and 1 division=10μm)
12. Keep sample slide under oil immersion and under same magnification.
13. Select at random one field and count the globules.
14. Find the number of divisions covered by these globules.
15. Take reading of about 200 globules.

❑❑❑

Chapter–2

Chemical Techniques for Quality Assessment

Laboratory Tests

Sampling procedures

Samples from number of containers- As soon as the cans are received at the reception dock, remove the lid, mix contents of each can with a plunger and draw equal quantities from each container and mix together. Draw approximately 80 – 100 ml of sample for testing. If the cans are of same diameter, Alternatively, milk from different containers are first poured into a vat and then the sample is drawn by thorough mixing as above..

Samples from storage tank – Before drawing sample from storage tank, switch on the mechanical agitator and allow the contents to mix for 15-20 minutes. Discard first 1.0 L of milk. Transfer 250-500 ml of the mixed milk from the outlet valve into a sample bottle.

Samples from rail / road tankers- Open the lid of the tanker and insert a big plunger through the main hole and move it forward, downward and backward bring the plunger to the surface at each stroke. Continue mixing for 15-20 minutes. Draw proportionate quantities of sample with the help of suitable dipper from different levels. Draw 250-500 ml of well-mixed sample into the sample bottle. The total volume of individual composite sample should not be less than 150 ml.

(i) Fat test

Aim :- To determine percentage of fat present in milk

(a) **Gerber method** :- When definite quantity of sulphuric acid and amyl alcohol are added to definite volume of milk, the proteins dissolve & fat globules are set free which remains in liquid state due to heat produced by the acid. Upon centrifugation, fat being lighter separates on top of the solution as liquid fat column in butyrometer stem.

Procedure

1. Transfer 10 ml of sulphuric acid (Gerber acid) into milk butyrometer (range 0 – 10%) using automatic measure.
2. Mix well the sample of milk and draw 10.75 ml with the help of milk pipette.
3. Transfer the milk slowly from the side of butyrometre wall taking care not to wet the neck of the butyrometer.
4. Add 1.0 ml of amyl alcohol using automatic measure. Clean & dry the neck of butyrometer with cotton if it is wet.
5. Stopper the butyrometer with a lock stopper, using the key, and shake the contents well.
6. Keep the butyrometer in water bath at 65° C for 5 mts.
7. Place the butyrometer in Gerber centrifuge, balance the machine and centrifuge the contents for 5 min. at 1100 – 1200 rpm.
8. Keep the butyrometer in water bath at 65° C for 5 mts.
9. Read the fat content (colour less column) directly on the butyrometer stem.

(b) **Gravimetric method** – For quantitative separation of fat it is necessary to break the protective film surrounding the fat globule by suitable agent (ammonia). The passage of fat globule from aqueous phase to solvent is facilitated by alcohol. The mixed solvent system brings about extraction of fat from non-fatty solids in the solution.

Requirements: - Rose-Gottleib tube / Majonnier flask, fat evaporating dish, rubber stoppers, ammonia solution, diethyl ether, petroleum ether, ethyl alcohol, centrifuge machine, hot plate, oven, desiccator, weighing balance (electronic-top loading)

Procedure

1. Weigh 10 g of well-mixed sample of milk into the extraction tube.
2. Add 1 ml of concentrated ammonia and mix well by gentle swirling motion.
3. Add 1 ml of alcohol and again mix well.
4. Add 25 ml of diethyl ether, stopper the tube with cork and shake vigorously for one minute.
5. Remove the cork and add 25 ml of petroleum ether, stopper and shake well for 30 seconds.
6. Allow the tube to stand until ethereal layer is separated clearly from the aqueous layer (for Rose-Gottlieb method)
7. In case of Majonnier method- centrifuge the Majonnier flask for 10 minutes.
8. Decant the clear ethereal layer into a previously weighed aluminum fat dish.
9. Repeat extraction with 15 ml portions each of diethyl ether and petroleum ether, twice to completely extract fat from the sample.
10. Transfer the ethereal layers from each extraction into the fat dish.
11. Place the fat dish on a hot plate (60° – 65° C) and allow the ether layer to evaporate slowly.

12. Dry the residual fat at 98° C in an oven and then allow the fat dish to cool by placing in a desiccator.
13. Weigh the fat dish and note the weight.
14. Repeat heating, cooling and weighing until successive weighing do not differ by more than 1 mg.
15. Calculate amount of fat by difference of weight in fat dish before and after extraction (reading A).
16. Divide reading A by the weight of sample and multiply by 100 to get percent fat in sample.

(c) Using Electronic milk tester –It is based on photometric measurement of light scattered by the homogenized milk sample having constant fat globule size to provide a fixed relationship between amounts of light scattered to the fat content. The intensity of scattered light passing through the cuvette (holding the sample) falls on photocell which converts it into electric current/ pulse fed to the digital read-out that directly corresponds to the fat content. The interfering milk proteins are dissolved by the chemical diluent.

Requirements: - Electronic milk tester, Di-sodium EDTA salt, di-sodium hydroxide, triton x-100 emulsifier, antifoam agent, saturated potassium dichromate solution (1%), pH meter, 10L plastic container.

Procedure

(a) **Preparation of diluent** - Weigh 45 g of EDTA and 7.6g of di-sodium hydroxide and dissolve in 2-3 L of distilled water. Add 0.5ml of triton x-100 and 0.5ml of antifoam agent. Mix the contents well and make-up the volume to 10L. Check pH (should be 9.5-10). Use this solution within two weeks.

(b) **Sample preparation** – Gently stir the milk and draw a representative sample. Test freshly drawn sample within 2-3 hours. Where sample needs to be preserved for 12 hours or more without refrigeration, add 1% saturated potassium dichromate solution. If samples are to be kept longer and transported, cool to 5-7°C.

(c) Installation and operation –

1. Install the EMT on a suitable platform for easy operation.
2. Place the diluent jar next to the EMT and connect the plastic pipe/tube to the inlet stub of EMT via a filter unit.
3. Connect a plastic tube to the out-let tub of the EMT and leave the other end into a waste container placed under the table.
4. Connect the EMT to electric line, switch on the instrument and allow to warm-up for at least one hour before measurement.
5. Place an empty beaker under the milk inlet tube and push "milk-in" and "milk-out" buttons alternatively until no air bubble is seen in the syringe.
6. Check the instrument for Zero check by placing empty beaker under milk in-let tube and press "milk inlet" and "milk outlet" button twice alternatively to fill the beaker with diluent.
7. Move the beaker to mix inlet tube and operate the handle for six times up and down.
8. Turn the zero knob to display 0.05 on the read out. Carry this check daily before use and then after every one hour during the day.
9. Gently mix the sample and place under milk inlet tube and press the "milk in" button.
10. Place a clean beaker under milk intake tube and press "milk out" button.
11. Move the beaker to mix intake tube.
12. Operate the homogenizer handle up and down three times.
13. Read the result on the digital display panel.
14. Repeat operation with out milk sample to clean the cuvette.

(ii) Solid-not fat test (s.n.f. test)

Aim :- To determine specific gravity of milk and calculate s.n.f. content of milk.

Lactometers are used for rapid determination of specific gravity of liquids. The method is based on the law of floatation which states that when a solid is immersed in a liquid it is subjected to upward thrust equal to the weight of liquid displaced by it and acting vertically upwards. Lactometers are variable immersion type hydrometers and calibrated before hand with liquid of known specific gravity.

Procedure

1. Adjust the temperature of milk sample to measuring temperature prescribed for lactometer (15.5° C).
2. Mix the sample well avoiding incorporation of air or foam formation.
3. Pour sufficient milk into the glass or steel cylinder to allow free floating of lactometer.
4. Place the lactometer in the milk and allow it to float till it stops and assumes a constant level.
5. Record the lactometer reading and temperature of milk at the same time.
6. Take another reading by flapping the top of the lactometer stem and when it again assumes constant level.
7. Take average of the two readings. Get corrected value (CLR) from the standard table for corresponding temperature.
8. Calculate the s.n.f. content using the given formulas.

Calculations :-

a) Specific gravity = CLR / 1000 + 1

b) Percent SNF = CLR / 4 + (0.2 x fat %) + 0.1

(iii) Total solid content (TS)

The total solid content of milk may be determined using both Richmond formula as well as gravimetric method. However, the second option is more reliable both for milk and milk products.

(a) Using Richmond formula

Aim :- To determine specific gravity of milk and calculate total solid content of milk.

Requirements: - Lactometer (ISI), lactometer jar, thermometer

Procedure

1. Adjust the temperature of milk sample to measuring temperature prescribed for lactometer (27° C).
2. Mix the sample well avoiding incorporation of air or foam formation.
3. Pour sufficient milk into the glass or steel cylinder to allow free floating of lactometer.
4. Place the lactometer in the milk and allow it to float till it stops and assumes a constant level.
5. Record the lactometer reading and temperature of milk at the same time.
6. Take another reading by flapping the top of the lactometer stem and when it again assumes constant level.
7. Take average of the two readings. Get corrected value (CLR) from the standard table for corresponding temperature.
8. Calculate the total solids (TS) content using the given formulas.

Observations: -

Sl.No.	Lactometer reading (actual)	Temperature(° C)	Corrected lactometer reading(CLR)
1.			
2.			
3.			

Calculations :- Percent Total solids = CLR / 4 + (1.25 x fat %) + 0.44

(b) Gravimetric method

Whole milk contains about 84 – 87 % water and remaining 13-16% are solids comprising of fat, protein, carbohydrate and minerals. By evaporating water from the milk under controlled conditions the total solid content can be determined accurately.

Aim :- To determine total solid content of milk gravimetrically.

Requirements: - Aluminium moisture dish, boiling water bath / hot plate, weighing balance, hot air oven, dessicator, tong.

Procedure

1. Weigh accurately a clean and dry empty dish. (A)
2. Transfer 5.0 ml of sample into the dish and note the weight. (B)
3. Place the dish on boiling water bath or hot plate for 20-30 minutes.
4. Allow the water to evaporate.
5. Place the dish in hot air oven at 100° C for 3 hours.
6. Transfer the dish to a desiccator and allow cooling for 30 minutes.
7. Weigh the contents and note the weight. (C)
8. Repeat heating and cooling until difference in two successive weights do not exceed 0.5 mg.

Observations

Weight of empty dish = A g

Weight of empty dish + sample = B g

Weight of dish after drying = C g

Calculations :-

Weight of sample = Reading B – Reading A

Weight of solids in milk = Reading C – Reading A

Percent total solids = C – A / B - A x 100

(iv) Moisture content

The procedure for determination of moisture content is similar to the procedure of total solid determination. Moisture content of sample can be determined by subtracting the value for total solids from 100.

Moisture percent = 100 – (% TS)

(v) Acidity Test

Normal acidity of milk is due to its constituents like casein, citrates, phosphates and CO_2. This acidity can be measured by titrating milk against a standard alkali solution using an indicator and is expressed in terms of lactic acid.

Aim :- to asses suitability of milk for heat processing.

Procedure

1. Transfer 10 ml of milk into a white porceline dish, with the help of pipette.
2. Add few drops of 1% phenolphthalein solution as indicator.
3. Titrate the contents with N/9 NaOH solution.
4. Observe occurance of pink colour as end point for the titration.
5. Note the titre value.
6. Calculate percent acidity of the sample as lactic acid.

Observation

Sl.No.	Initial reading	Final reading	Vol. Of alkali used	% acidity
1.				
2.				
3.				
4.				

Calculation: - Titrable acidity % (as lactic acid) = 9x V1 x N / V2

Where,	V1	=	Volume of N/10 NaOH used
	V2	=	Volume of milk sample
	N	=	Normality of NaOH used

(vi) Protein content

When nitrogenous organic substance like protein is digested with acid the organic nitrogen is converted to inorganic nitrogen as ammonium sulphate. Upon distillation with alkali ammonia is liberated and absorbed in standard acid. Excess acid is titrated with standard alkali to colourless end point. The test estimates total nitrogen which when multiplied by the factor 6.38 will give total protein content of milk.

Aim :- To determine protein content of milk by Kjeldahl method.

Requirements: - Kjeldahl flask, round bottom flask, conical flask, burette, digestion heater, distillation unit, conc. sulphuric acid, copper sulphate, potassium sulphate, N/10 NaOH, 50% NaOH, N/10 sulphuric acid, mixed indicator (methyl red and 0.1% methylene blue in ethanol).

Procedure

1. Weigh accurately 10 g milk in a clean and dry Kjeldahl flask.
2. Add 25 ml of concentrated sulphuric acid, 0.2 g of copper sulphate and 10 g of potassium sulphate.
3. Place the flask on digestion heater under fume chamber and heat gently to boil until contents are clear, and then for another 2 hours.
4. Allow the liquid to cool and dilute with 200 ml of distilled water.
5. Fit the flask with micro-kjeldahl distillation unit.
6. Carefully add 75-80 ml of 50% NaOH down the neck of the flask to form a layer under the acid.

7. Take 50 ml of N/10 sulphuric acid in a 250 ml conical flask and place it under the condenser so that the tip of the condenser is dipped in the liquid.
8. Start heating and distil the contents until all ammonia has passed over as indicated by collection of 150 ml distillate.
9. Stop heating and remove the conical flask. Wash the condenser tip to remove all traces of condensate.
10. Titrate the contents against N/10 NaOH using mixed indicator.
11. Carry out blank determination by using 0.5 g pure sucrose in place of milk.

Observations: - Record the titer volume for original and blank experiments in tabular form.

Sl.No.	Initial reading	Final reading	Vol. Of alkali used
1.			
2.			
3.			
4.			

Calculation: - Percent total nitrogen by weight = 1.4 (B-A) N / W

Percent total protein = Percent total nitrogen x 6.38

Where B = Volume of N/10 NaoH used for blank determination.

A = Volume of N/10 NaOH used for sample.

N = Normality of standard NaOH.

W = Weight of sample in g.

(vii) Lactose content

Lactose solution reduces the cupric salt solution, quantitatively, to cuprous oxide upon boiling. The quantity of lactose can be calculated from the amount of copper salt reduced.

Aim :- To determine lactose content of milk by Lane-Eynon method.

Requirements: - Volumetric flask, beaker, conical flask, burette, funnel, filter stand, chemical balance, hot plate, filter paper, 10% acetic acid, Standard Fehling A and B solution, 1% methylene blue in water, 0.5% lactose solution.

Standard lactose solution – Dissolve 1.25 g of lactose powder in 250 ml distilled water.

Procedure

1. Take 25 ml of milk into 500 ml conical flask and dilute to 200 ml with distilled water.
2. Add 3.75 ml of 10% acetic acid solution and boil.
3. Cool the contents and transfer to 250 ml volumetric flask and make up the volume with distilled water.
4. Filter through No. 42 filter paper and collect the filtrate in a clean dry beaker.
5. Fill the burette with this filtrate. This contains lactose.
6. Pipette 5 ml each of Fehling A and B solution into 250 ml conical flask.
7. Heat the contents to boiling, add 5 drops of indicator and titrate with filtrate from burette 1 ml at a time till blue colour changes to red.
8. Note the volume of filtrate used for titration.
9. Pipette 5 ml each of Fehling A and B solution into another 250 ml conical flask, and add whole volume of filtrate used in step 7 so that not more than 0.5-1.0 ml will be required to complete titration.
10. Heat the mixture to boiling, add 5 drops of indicator and titrate with filtrate to a brick red colour end point.
11. Repeat titration to get values within 0.1 ml.
12. Run a blank using 5 ml each of Fehling A and B solutions and titrate against 0.5% lactose solution.
13. Note the volume of standard lactose solution used and calculate amount (mg) of lactose present.

14. Similarly note volume of milk filtrate used in titration and calculate the amount of lactose present in milk sample.

Observations: - Record the titer volume for original and blank experiments in tabular form. Make separate tables for blank and original titration values.

Sl.No.	Initial reading	Final reading	Vol. Of filtrate used
1.			
2.			
3.			
4.			

Titer value (blank) = Volume of standard lactose solution used for titration.

Titer value (sample) = Volume of milk filtrate used for titration

Calculation :-

Lactose (mg) in standard solution (A) = Titer value (blank) × 0.5 / 100

Lactose (mg) in sample = A × 250 / titer value (sample)

Lactose percent in sample = (A × 250 / titer value × 25) 100

(viii) Total Ash content

Milk contains soluble salts like sodium, potassium, calcium, phosphorus, citrates, sulphates, chlorides, carbonates, magnesium etc. Heating milk at higher temperatures decomposes organic matter and the soluble inorganic salts are left in the form of ash.

Aim :- To determine total ash content of milk.

Requirements: - Single pan balance, muffle furnace, desiccator, silica crucible, tong, hot plate.

Procedure

1. Weigh the empty and dried silica crucible and note the weight.
2. Accurately weigh 10 g if milk sample and note the weight.

3. Evaporate the sample to dryness on a hot plate.
4. Place the crucible in a pre-heated muffle furnace and heat the contents at 550° C until ash is free from carbon.
5. Cool the crucible by placing in desiccator.
6. Weight the crucible containing the ash.
7. Continue heating, cooling and weighing until the difference in weight is not more than 0.1g.

Observations: - Record the weights.

a) weight of empty crucible = W g
b) weight of crucible with milk = W1 g
c) weight of crucible after drying = W2 g
d) weight of milk sample = W1 – W g (A)
e) weight of ash = W2 – W g (B)

Calculation: - Percent ash by weight = B / A × 100

(ix) Colorimetric determination of Urea

Urea is a common constituent of blood and other body fluids and is normally present in milk, as part of NPN, in the range of 100 – 450 ppm. The test is not affected by chilling, pasteurisation, addition of adulterants and nutralizers.

Requirement:- Test tube, volumetric flask, filter paper, beakers, DMAB reagent, Phosphate buffer (pH 7.0), TCA, colorimeter / spectrophotometer.

DMAB reagent– Dissolve 1.6 g of p-dimethy lamino benzal dehyde in 100ml ethyl alcohol containing 10% concentrated HCl.

Standard Urea solution – Weigh 100mg of AR quality Urea and dissolve in phosphate buffer and make up volume to 100ml in a volumetric flask. This will be equivalent to 1mg / ml urea .

Phosphate buffer – Weigh 3.40g of anhydrous potassium dihydrogen orthophosphate and 4.35g dipottasium monohydrogen orthophosphate in 100ml distilled water.

Diluent solution – Mix equal volume of 24% TCA and phosphate buffer.

Reagent blank – Mix equal volumes of diluent solution and DMAB reagent.

Preparation of standard curve

1. Measure different concentrations of standard urea solution (0.1mg to 2.0mg) in different test tubes.
2. Make total volume of 5ml with diluent solution.
3. Add 5ml of DMAB reagent and mix the contents well.
4. Measure optical density of each tube at 425 nm
5. Plot a graph of optical density against concentration to obtain standard curve.

Estimation

1. Mix 10ml of milk sample with 10ml of 24% TCA.
2. Filter the contents using Whatman no.1 filter paper and collect filtrate. Discard first few drops.
3. To 5ml of filtrate add 5ml DMAB reagent and mix well.
4. Observe optical density at 425 nm against reagent blank.

Result - Calculate the amount of Urea from standard curve.

(x) Presence of Vegetable fat

Aim :- To detect presence of vegetable fat in milk

Vegetable fat being cheaper than milk fat is used as an adulterant as it is easily mixed and homogenized into the milk. butyro-refractometer (B.R) reading of pure milk fat is between 40-43.

Procedure

1. Transfer 100 ml of milk into a 500 ml separating funnel, add 10 ml of concentrated Nitric acid and mix well.
2. Add 50 ml alcohol and 100 ml each of diethyl ether and petroleum spirit and mix the contents by shaking gently for 2-3 minutes.

3. Keep separating funnel un-disturbed for 20-30 minutes.
4. Discard the bottom aqueous layer and collect ether layer in 500 ml conical flask.
5. Evaporate the ether on a boiling water bath.
6. Place the flask in electric oven at 120° C for 15-20 minutes.
7. Apply a drop of melted fat on prism of butyro-refractometer maintained at 40°C by hot water circulation.
8. Note the butyro-refractometer (BR) reading.

Opacity test

1. Take 10 g of isolated fat in a test tube.
2. Keep in water bath at 50° C for 3-4 minutes.
3. Transfer the test tube at 23° C and hold for 15-20 minutes.
4. Note the time for appearance of turbidity (opacity time).
5. Opacity time for pure milk fat is 14-15 minutes.
6. Refined oil has higher opacity time (19-20 minutes).

(xi) Detection of buffalo milk in cow milk

The test is known as Hansa Test which detects admixture of buffalo milk in cow milk. This is based on the principle of antigen-antibody reaction. The antiserum produced after immunization of rabbit with buffalo casein, gives visible reaction only with buffalo milk.

Requirements : Rabbit (adult healthy male), autoclave, centrifuge, syringes, tubes, pipettes, spirit, xylene, saline, adrenaline(1:50,000 IU), pure cow and buffalo milk.

Procedure

(a) Preparation of antigen

1. Take 50 ml of buffalo milk in a PP tube and centrifuge at 3000 rpm for 30 minutes.
2. Break the fat layer and decant skim milk in a beaker.

3. Centrifuge the skim milk at 15,000 rpm for 30 minutes.
4. Remove clear whey and scrap casein micelles in a beaker containing distilled water.
5. Homogenize and makeup volume equal to volume of skim milk used.
6. Filter and add 5% phenol solution @ 3% of serum volume to preserve, and store in refrigerator till further use.

(b) Immunization

1. Select healthy male rabbit and inject antigen as per schedule.
2. Inject intra-peritonial fluid taking care not to pierce the viscera.
3. Administer intra-venous injections, in marginal vain of the ear, slowly.

(c) Blood collection / testing

1. Collect blood from each rabbit on first day of third weak to get serum.
2. Separately, dilute cow and buffalo milk 1:10 with water.
3. Place a drop of diluted milk on a clean slide, add one drop of serum to be tested and mix well with toothpick.
4. Observe agglutination.
5. Further immunization is stopped for rabbits whose serum gives good titer and blood collected to maximum extent to get serum.
6. For other rabbits, the immunization continues till they give good titer. Suspend immunization after 6 weeks.
7. Agglutination reaction shows presence of buffalo milk in the sample.
8. Clear solution confirms pure cow milk.

Immunization schedule

Immunization weak	Dosage of antigen (ml) per weak		
	Day 1	Day 2	Day 3
I	0.5	0.5	1.0
II	1.5	1.0	1.5
III	2.0	1.5	2.0
IV	3.0	2.0	2.5
V	4.0	2.5	3.0
VI	5.0	2.5	3.0

(xii) Detection of abnormal milk

The variation in the composition of milk by physical, chemical and microbiological sources make it abnormal. The problems posed by such milk are related to its processing and to consumers causing health hazards. It is therefore necessary to detect abnormalities and rectify the same before use. Mastitis, a memory gland inflammation, manifest physical and chemical changes in milk properties the major being change in the chloride content of milk. The disease may also cause presence of excessive number of leucocytes and mucosal epithelial cells.

Requirements : Strip cup, Microscope, Breeds slide, Breeds pipette, methylene blue chloride, ethyl alcohol, tetrachloroethane, glacial acetic acid etc.

Procedure

(a) Strip cup test

It is a simple field test to detect presence of mucosa and milk clots in foremilk.

1. Take first 2-3 strips of fore milk onto a fine mesh cloth or over a cup
2. Observe for any clots, mucosa or blood cells on the screen.
3. Presence of clots, mucosa or blood cells indicate infection.

(b) Leukocyte counts (Microscopic test)

1. Transfer 0.01 ml of milk sample on Breeds slide and spread on marked area.
2. After air drying, dip the slide in Newman stain for 30-60 seconds.
3. Dry in air, mount and observe under oil immersion lens and observe for presence of cells (WBC).
4. Calculate number of WBC / ml of sample. Normal milk contains 70,000-1,00,000 WBC /ml
5. Counts of over 1,50,000 indicate infection.

(c) Bromthymol blue test

1. Transfer 5.0ml of freshly drawn milk in a test tube.
2. Add 1.0ml of 0.04% aqueous solution of bromthymol blue
3. Mix well and observe colour
4. Presence of greenish blue or blue colour indicates mastitis infection.

(xiii) To determine the casein content in milk by Walker-Formol method

Principle - The casein content of milk represents about 80% of milk proteins. Casein plays an important role in cheese making. The quantity of casein in the milk determines the yield as well as the uniformity of quality of the cheese being manufactured from the milk. Potassium oxalate releases the calcium ions from the protein micelles and brings the micelles into ionic form. Formalin is added to liberate hydrogen ions from casein. These ions are then titrated with NaOH.

Requirements :- Saturated potassium oxalate solution, phenolphthalein indicator, N/10 NaOH, neutral formalin

Procedure

1. To 10 ml milk, add 0.4 ml saturated potassium oxalate solution and 1 ml phenolphthalein indicator.

2. Wait for 5 min.
3. Titrate with N/10 NaOH solution.
4. To this add 2 ml neutral formalin.
5. Wait 5 min again.
6. Titrate against N/10 NaOH.
7. Note the second titer reading.

Observations & Calculations

Trial	Second titration: Initial burette reading (a)	Second titration:Final burette reading (b)	Second titration: Titer value (b-a)
1			
2			
3			
Mean			

Per cent casein = d x 1.34

Per cent Protein = d x 1.70

(xiv) Detection of pesticide residues

The status of chemical pesticide has undergone a sea change from green revolution to an agent polluting air, water and soil and posing serious threat to public health by entering into food chain. The danger is encountered from organochloro pesticide residues (OCPR) like DDT, BHC etc. since these are fat soluble, persistent, non-degradable and also find easy transfer to milk through animal feeds, concentrate and environment.

Requirements : HPLC equipped with SPD-M10 AV/UV or ECD detector, Reference standards of DDT, BHC, Pet. Ether 40-60°C, Dichloromethane, Florisil (PR- Grade), Glass Columns (22 mm I.D. X 600 mm)

(a) Isolation Of OCPR By Column Chromatography

Principle

Analysis of OCPR in milk and milk products essentially by

IDF standard (1991) is based upon adsorption clean-up. This involves mixing of test portion in presence of water with florisil until a homogenous powder is obtained, transfer of this mixture to florisil column, selective elution of pesticides and concentration of the elute.

Procedure

1. Take 100g of florisil (60-100 mesh) in a beaker and wash 3-4 times with 200 ml portions of petroleum ether.
2. Air dry and activate at 600°C for 12-14 hr.
3. Weigh 50g florisil and deactivate by equilibrating in 3% water for 20 minutes with intermittent shaking.
4. Add 100 ml pet. ether in 22 mm X 600 mm glass column and pack with 25 g deactivated florisil (bottom layer).
5. Take 25 g activated florisil and mix well with the following test portions (top layer) as the case may be:
6. Take 5 g milk of liquid milk, or reconstituted 10 g skim milk powder and 1 g of whole milk powder/infant milk food with 10 ml of HPLC grade water. Dissolve 0.25 g fat/ghee in 2.5 ml of petroleum ether 40-60°C.
7. Transfer the sample quantitatively to the top of the column containing 25 g deactivated florisil.
8. Elute sample with mixture of Pet. Ether : Dichloromethane (4:1) at a flow rate of 1-2 ml/min.
9. Collect eluent in a clean beaker and dry under nitrogen.
10. Dissolve residue in 100 µl of HPLC grade methanol.
11. Simultaneously also run a blank experiment

(b) Estimation of Individual OCPR

(i) By HPLC method

Principle

Analysis of OCPR is done over reverse phase liquid

chromatographic (RPLC) octadecylsilyl (ODS) column in which ODS groups are bonded to the silica surface. Solute retention is mainly due to hydrophobic interactions between the solutes and the hydro carbonaceous stationary phase surface. Polar mobile phase elutes the solute in order of decreasing polarity (increasing hydrophobicity) and increasing the polar (aqueous) component of the mobile phase increases the retention of the solutes. UV-200 detector detects the fractionated peaks signals.

Procedure

1. Fix a 250 mm x 4 mm ID octadecylsilyl ODS column in a reverse phase HPLC equipment fitted with UV detector.
2. Attach the HPLC with a professional computer and printer.
3. Inject 2-20 μl of standard reference mixture using micro pipette.
4. Elute mixture at 50° C using mobile phase of methanol-water (80:20) at a flow rate of 1 ml per minute.
5. Note retention time and peak area of reference standard.
6. Similarly, inject sample obtained form florisil column using similar conditions as for reference standard.
7. Identify and estimate individual OCPR by comparing retention time and peak area to those of reference standards.
8. Express concentration of pesticide as μg per gram sample.

RESULT: The individual OCPR are identified and estimated through retention time data and peak area data in comparison to those of reference standards.

Conc. of pesticide (mg/g milk) = $C \times a_2 \times V / a_1 \times V_2 \times W$

where,

C = Conc. of reference standard (mg)

a_1 = Area of reference standard

a_2 = Area of sample peak

V_2 = Volume injected (ml) of sample

V = Total volume (ml) of sample

W = Wt. of milk (g)

(ii) By TLC method

Procedure

1. Take 20 x 20 cm sheet of alufoil precoated with 0.2 mm thick neutral aluminium oxide layer.
2. Load the sheet with 20 μl sample isolated from florisil column.
3. Also load 20 μl blank solution and 1-5 μl reference standard.
4. Place sheet into TLC chamber filled with cyclohexane and leave undisturbed for one hour at 25° C.
5. Remove foil and spray with 0.5% alcoholic silver nitrate solution.
6. Allow the reaction to occur for 10 minutes.
7. Irradiate sheet under UV source for 10 minutes and spray with distilled water.
8. Again expose to UV irradiation for 1-2 minutes for appearance of violet black spots.
9. In case spots are not clearly visible, irradiate for another 10 minutes.
10. Moisten and continue irradiation until spot for lowest concentration of standard is visible.

❑❑❑

Chapter–3

Microbiological Techniques for Quality Assessment

In the last decade or so, the problem of determining the microbial content and their activities in dairy products has been under close scrutiny. The dairy industries still adopt the dye reduction test because it is cheap, simple and satisfactory. However, other new techniques have been introduced along with some modifications in traditional tests. The microbiological quality must be maintained. To assess the bacterial content and their activities in dairy products, various methods are available for the microbiological testing of milk. The specific tests should be chosen as per the application and other factors like cost etc. On the basis of public health and economic consideration it is necessary to test microbiological quality of milk at a number of points along the chain from producer to consumer which includes - raw milk supply, bulk milk arriving at processors premises, monitoring bulk milk during storage prior to processing and monitoring the procession & the quality control of finished products. The techniques for estimating bacterial numbers in raw material are divided into two categories.

Sampling procedures: All equipments used should be cleaned and sterilized properly by any of the following methods –

1. Exposure to hot air at 170°-175° C for 2 hours.
2. Exposure to steam at 121° C for 20 minutes.
3. Immersion in 70% ethanol solution.
4. Ignition with 96% ethanol.
5. Exposure to sufficient dose of Gamma radiation.

Using aseptic conditions draw a representative sample with sterilized equipment and store in sterilized container. Such samples should be immediately chilled to 4° C and maintained at this temperature until analysed. The interval between sampling and analysis should not exceed 4 hours. Time of sampling and analysis should be recorded on the report. Samples for microbiological analysis should be drawn by trained persons.

Raw Milk – Clean container from outside and wipe-off the mouth of container with cloth dipped in 70% alcohol. Mix the contents thoroughly by inverting or stirring using sterilized plunger. Do not raise plunger above surface of liquid. Draw 100ml of sample aseptically and transfer into sterile bottle.

Pasteurized Milk – Clean exterior of pouch with tap water and dry by wiping with a clean cloth. Wipe a corner of the pouch with cloth soaked in alcohol. Sterilize a scissor by dipping in alcohol followed by flaming to evaporate alcohol. Cut the sterilized corner and collect 100ml sample in a clean and sterilized bottle.

Make appropriate dilution by transferring 1ml milk sample into 99ml ringer's solution before plating. Make further dilutions if required.

Cream – Mix cream with sterilized stirrer. If frozen it may be tempered at 40° C and mixed thoroughly. Transfer 100ml cream aseptically into sterile bottle. Transfer 1ml of well mixed cream sample into 9ml or 99ml sterile diluent.

Evaporated / Condensed milk – Let container stand at 45° C for 30 minutes in water bath. Remove and clean the container, mix the contents well for 5-8 minutes using broad bladed stirrer with one edge shaped to the contour of container. Draw sample (100g) aseptically from different portion of container and transfer to sterile bottle.

Ice-Cream – Transfer 11g melted sample directly into 99ml sterile diluent.

Butter – Using sterile knife / spoon / spatula remove surface layer from center of container to a depth of 5mm. Take sample from different portions using sterile butter trier. And transfer quickly in sterile bottle. Place bottle in water bath at 45° C for 15 minutes, mix well and draw 1ml melted butter and pour into 99ml of sterile and warm diluent to get 1:100 dilution.

Panner – Weigh 50g panner and grind to fine paste in sterile grinder jar. Take 10g of prepared sample into flask containing 90ml sterile diluent (primary dilution). Transfer 1ml of this solution into 9ml sterile diluent.

Cheese – Disinfect the cheese surface with undenatured ethanol. Insert cheese trier to a depth of 25mm and rotate full turn and withdraw with the core. Similarly, withdraw sample with smaller trier from the same hole. Transfer content to sterile bottle. Repeat until 100g sample is collected. Proceed further as in case of panner.

Dry Milk Powder – Take sample with dry and sterile borer from near the center of container, by passing borer through the product at even rate of penetration. Rotate through 180 degree and withdraw. Discharge contents in clean, dry and sterile bottle.

Microbiological tests for raw milk

Dye reduction tests

The test is based on dehydrogenase activity of bacteria present in milk. As bacteria multiply they consume dissolved oxygen resulting in decrease of radox potential of milk. The rate of reduction of dye is an index of bacterial number present in milk. The reduction time is directly proportional to the bacterial load of the sample.

(i) Resazurin Test (10 minutes)

Aim :- To assess hygienic / bacteriological quality of milk in terms of its effect on rate of acid development and grading of milk.

Stock solution :- Dissolve 0.05 g resazurine powder in 100 ml of distilled water. Heat for 30 minutes and cool.

Test solution :- Transfer 1.0 ml of stock solution into 10.0 ml volumetric flask and make-up the volume with distilled water.

Procedure

1. Transfer 10 ml of milk in a sterile test tube aseptically.
2. Add 1 ml of Resazurine Test solution.
3. Stopper the tube and shake well.
4. Place tube in a water bath at 37° C for 10 minutes.
5. Compare the colour developed in a Lovibond comparator using Resazurine disc.
6. Grade the milk as per ISI Standard.

ISI Standard for resazurine test

Disc number	Milk Quality	Remarks
4 and higher	Satisfactory	Milk accepted
3.5 – 1.0	Doubtful	Perform other tests
0.5 – 0.0	Unsatisfactory	Reject milk

Redox potential	Reaction product	Colour
0.3 volts	Resazurin	Purple blue
0.2 volts	Resafurin	Pink
0.05 volts	Dihydroresafurin	White

(ii) Methylene blue reduction test (MBR) 30 minutes

Aim :- To assess bacteriological quality of milk

Stock solution – Dissolve one tablet of methylene blue in 200 ml of distilled water.

Working solution – Dilute one part of stock solution in three parts of distilled water

Procedure

1. Take 10 ml of milk in a test tube.

2. Add 1 ml of 1% methylene blue solution (working solution)
3. Stopper the tube and shake well.
4. Place the tube in a water bath at 37° C.
5. Observe discolouration of the contents against control sample.
6. Note the time of discolouration.
7. Make control sample by mixing 10 ml of milk + 1 ml of water and 10 ml milk + 1 ml methylene blue dye. Keep these in boiling water bath for 3 minutes.
8. Samples with less than 30 minutes discolouration time is not suitable for acceptance.

Standard for methylene blue test

MBR time	Milk quality
5 hours and above	Very good
3 – 4 hours	Good
1 -2 hours	Fair
½ hours and less	Poor

(iii) Direct Microscopic count

This method consists of examining a stained film of milk under oil immersion lens of a compound microscope. This method is less expensive and rapid for determination of extend of bacterial contamination in milk.

Requirements :- Microscopic slide (2.5 x 7.5 cm), stage micrometer, compound microscope, Breeds pipette, forceps, staining tray/jar.

Preparation of methylene blue stain :- Take 54 ml of ethyl alcohol in 40 ml tetrachloroethene in a 150 ml conical flask and heat the contents to 70° C to allow mixing of two liquids.. Weigh 1 g of methylene blue powder in a separate dry container. Add warm mixture to methylene blue powder and shake vigorously until dye is completely dissolved. Cool the contents and slowly add 6 ml of glacial acetic acid and mix. Filter and store in tightly stoppered bottle.

Determination of microscopic factor :- Place the stage micrometer and focus with 10 X ocular lens. Add a drop of immersion oil on the micrometer and focus under oil immersion lens. Count the number of small spaces (0.01mm) in the diameter of the field ranging from 0.146 to 0.206 mm. Calculate microscopic factor using the formula - **MF = 10,000 / 3.1416 x r^2.**

The reciprocal of MF represent fraction of 1ml of milk actually observed under one microscopic field. Thus the total count per ml milk will be N x MF. Where N = average number of bacteria in one microscopic field.

Procedure

1. Calibrate the microscope and determine microscopic factor (MF) with the help of stage micrometer under oil immersion lens.
2. Prepare a slide by drawing 0.01 ml milk using Breeds pipette and spread over marked area of 1 sq.cm with a thin wire needle.
3. Allow the milk film to air dry.
4. Place enough stain (1% Methylene blue) on the slide using a dropper, place it in slanting position and allow it to dry.
5. Wash the slide with water to remove excess stain and allow to dry.
6. Add a drop of immersion oil on the slide and using oil immersion lens examine the bacteria and count numbers in several fields.
7. Determine average number of bacteria per field (5-60).
8. Multiply the number of bacteria per field by the microscopic factor to obtain number of organisms per ml milk.
9. Compare result with following standard to categorize milk quality.

BIS standard for bacteriological quality of milk

Counts per ml	Bacteriological quality
Less than 500,000	Good
500,001 – 4,000,000	Fair
4,000,001 – 20,000,000	Poor
Over 20,000,0000	Very poor

(iv) Detection of inhibitory substances (antibiotic)

Screening test

1. Prepare inhibitor free reconstituted skim milk to serve as control
2. Place 10ml of raw milk sample in a sterile test tube, stopper and pasteurize the contents at 80° C for 5 minutes.
3. Take 10ml of reconstituted skim milk and treat in similar manner.
4. Dilute active starter culture (St.thermophillus) in sterile skim milk in the ratio of 1:1.
5. Add 1ml of diluted starter to each tube aseptically and incubate at 36° C for 2 hours.
6. Add 0.3ml of 1:25 aqueous solution of 2,3,5-triphenyl tetrazolium chloride reagent and further incubate for 30 minutes.
7. Remove the tube and compare intensity of red colour with control.
8. A lighter shade of unknown sample, than control, indicates presence of inhibitory substances.
9. These can be antibiotic / sanitizing agent (chlorine) / high leucocytes count or phage.

Confirmatory test for penicillin

Requirements

Antibiotic assay agar (Modified bacto PM indicator agar)

consisting of 3.0g beef extract; 5.0g Peptone; 3.0g dextrose; 1.7g tryptone; 0.3g soytone; 1.0ml polysorbate-80 and 10.0g Oxide agar No.1. Dissolve contents in 1000ml distilled water and adjust pH to 7.5 and sterilize.

Penicillin solution – (a) Stock : Aqueous solution containing 2000 IU/ml.

(b) Serial solution: Using stock solution, make serial solution to contain penicillin ranging from 0.004 IU/ml to 1000 IU/ml.

Spore preparation

1. Inoculate active culture of B.sterothermophillus @ 1% into 100ml aliquot of sporulating medium into 500ml flask.
2. Incubate at 55° C on a rotary shaker at 120 rpm till maximum sporulation occurs as seen by staining.
3. Centrifuge at 3000 rpm for 30 minutes to obtain spores.
4. Suspend spores in chilled sterile distilled water.
5. Wash for 4 times and finally centrifuge at 4000 rpm for 120 minutes.
6. Store at 5° C till use.

Assay

1. Adjust spore concentration to 50% transmission.
2. Add 2ml of this solution to 100ml liquefied antibiotic assay agar.
3. Pour this seeded agar (15ml) in petri dish and allow to solidify.
4. Make 6 equal bore wells of 8mm dia using s.s. cork borer.

Standard curve

1. Measure 0.05ml each of serial dilution, of penicillin, and deliver in seeded agar wells.
2. Incubate at 55° C for 2.5 hours.
3. Measure zone inhibition accurately with calipers.

4. Plot the values on semi log paper
5. A straight line is obtained using regression equation.

Detection in milk

1. Add 0.5 units of penicillinase enzyme solution into test tubes containing suspected milk.
2. Mix well and incubate at 37° C for 15 minutes.
3. Deliver 0.05ml of treated milk sample in seeded agar well.
4. To another well add control sample containing 0.008 IU/ml of penicillin in sterilized skim milk.
5. Incubate at 55° C for 2.5 hours and measure inhibition zone.
6. Calculate concentration of penicillin in unknown from standard curve.
7. Report as units/ml after multiplying with 20.

Confirmatory test for streptomycin

Streptomycin assay agar - Weigh 3.0g Yeast extract; 1.5g beef extract; 6.0g Peptone and 15.0g agar. Dissolve contents in 1000ml distilled water and adjust pH to 7.9 and sterilize.

Streptomycin solution – (a) Stock : Aqueous solution containing 4000 µg/ml.

(b) Serial solution: Using stock solution, make serial solution to contain streptomycin ranging from 0.01ug/ml to 100 µg/ml. Use this solution for making standard calibration curve.

Assay – Boil the test milk for 5 minutes to destroy any natural inhibitor. Then process as in case of penicillin.

Alternative tests

Test-I

1. Boil 1.0ml of milk with 1.0ml of N/10 NaOH for few minutes.
2. Add few drops of 1% HCl followed by few drops of 10% ferric chloride solution.

3. Development of brilliant violet colour confirms presence of streptomycin.

Test-II

1. Add 1.0ml of 2N NaOH into milk sample.
2. Place in boiling water bath for 3 minutes.
3. Cool to 10° C and add 1.0ml of phenol reagent.
4. Allow to stand for few minutes then add 3.0ml of 20% sodium carbonate.
5. Record development of colour both qualitatively and quantitatively.

(v) Rapid test for beta-lactam antibiotic using betastar kit

The beta lactam class of antibiotics are - PenicillinG, Ampicillin, Amoxicillin, Cloxacillin, Oxacillin, Nafcillin, Cephacillin etc. These are extensively used for treatment of mastitis in dairy animals. They may transmit to milk as contaminants. This is a receptor assay test involving specific beta lactam receptor linked with gold particles. The milk containing antibiotic is incubated with receptor an interaction between sample and receptor takes place forming the first band. The solution is then transformed into an immunochromatographic medium that appears as red colour reference band on the test strip. The colour intensity of test band is compared with reference band on the given chart.

Procedure

1. Remove kit from refrigerator and keep for 10 minutes at room temperature.
2. Open the vial and carefully transfer 1350 µl of previously boiled and cooled distilled water into the vail.
3. Mix the contents to completely dissolve contents of vail.
4. Add 25 µl of solution into eppendorf tube followed by 100 µl of suspected milk sample.

5. Mix and incubate at 41.5 °C for three minutes.
6. Dip the test strip into incubated vail keep the arrow side downwards and incubate for further two minutesRemove the strip and observe band formation and interpret the results as follows -
    a) No red colour – invalid test.
    b) First band Intensity more than reference band – test considered negative.
    c) First band Intensity lower than reference band – test considered Positive.
    d) In absence of first band – test is considered highly Positive

Note : In case of doubt the test should be confirmed by microbiological assay.

(vi) Rapid test for Tetracycline using Tetrasensor kit

This is a receptor based assay for rapid determination for presence of tetracycline molecule in milk. The test requires use of two elements, a reagent containing labeled receptor and a dipstick with two membrane capture lines that turns from green to red. The limit of detection is around 25 ppb.

Requirments :- Tetrasensor kit with micro well tray, reagent with labeled receptor, dipsticks, pipette, pipette tips.

Procedure

1. Place the empty microwell tray on the table.
2. Open the reagent tube and place reagent into as many microwells as number of samples for testing.
3. Add 200 l milk sample into microwells containg the reagent.
4. Mix to get a homogenous pink colour and incubate at room temperature for three minutes.
5. Dip an stick into the well and incubate for seven minutes at room temperature.

6. Remove the stick and compare colour lines on the strip.
7. For optical measurement use readsensor.
8. Draw a graph between readsensor reading and ng/ml tetracycline concentration for control sample
9. Compare reading of test samples by plotting on graph.

Result interpretation (visual comparision of colour intensity)

a) When no colour line occurs – the result is non – valid
b) When test line is prominent than control line – the test is Negative (< 25 ppb).
c) When test line is less visible than control line – the test is Positive (> 25ppb)
d) Absence of test line indicates highly positive sample.

(vii) Total plate count / Total viable count (Plate method)

This method consists of mixing a definite volume of milk with melted nutrient agar, solidification of agar and counting of colonies that appear after incubation. The result is reported as number of colonies per ml of milk.

Requirements :- Dilution water, Plate count agar, Conical flask, Pipettes, Water bath, Incubator, Petri dishes, test tubes, dilution bottles etc.

Preparation of Phosphate buffer :- ***Stock solution*** – Weigh 34 g of potassium dihydrogen phosphate in a clean and dry container and add 500 ml distilled water. Adjust pH 7.2 with 1N NaOH. Transfer the contents in 1000ml volumetric flask and makeup the volume.

Working solution – Take 1.25 ml of stock solution in 1000ml volumetric flask and makeup the volume. Sterilize the contents in 100ml bottles before use.

Preparation of Ringers solution :- ***Stock solution*** – Weigh 9.0 g of sodium chloride, 0.42 g of potassium chloride, 0.48g of calcium chloride and 0.2g of sodium hydro chloride in a clean and dry container. Transfer the contents in 1000ml volumetric flask and makeup the volume.

Working solution – Take 250 ml of stock solution in 1000ml volumetric flask and makeup the volume. Transfer the contents (99ml) in several dilution bottles or 10ml in series of test tubes. Stopper and sterilize the contents.

Procedure

1. Transfer 1.0 ml or 1.0 g of well mixed sample in a test tube.
2. Add 9.0 ml of freshly prepared Ringers solution (phosphate buffer, pH 7.2) to get a dilution of 1:10.
3. Transfer 1.0 ml of 1:10 solution to subsequent tubes and makeup the volume to get series of higher dilutions.
4. Transfer 1.0ml of required dilution of the sample, in duplicate, to sterile Petri dish.
5. Introduce 10-15 ml of melted media (45° C) in each plate under aseptic conditions.
6. Cover the plates and gently swirl the contents to form thin layer of media with test portion.
7. Allow the medium to solidify and then transfer the plates to incubator and incubate the plates at 37° C for 48 hours.
8. Remove the plates and count the colonies formed with the help of colony counter.
9. Count number of average colonies per plate and multiply with the dilution factor.
10. Compute the count per ml or per g and report.

(viii) Yeast and Mould count

Aim :- To estimate presence of yeast and mould in milk and milk products.

Requirements :- Dilution water, Potato dextrose agar, Conical flask, Pipettes, Water bath, Incubator, Petri dishes, test tubes, dilution bottles etc.

Procedure

1. Transfer 1.0 ml or 1.0 g of well mixed sample in a test tube.
2. Add 9.0 ml of freshly prepared Ringers solution (phosphate buffer, pH 7.2) to get a dilution of 1:10.
3. Transfer 1.0 ml of 1:10 solution to subsequent tubes and makeup the volume to get series of higher dilutions.
4. Transfer 1.0ml of required dilution of the sample, in duplicate, to sterile Petri dish.
5. Adjust pH, of required amount of media, to 3.5 with sterile 10% tartaric acid.
6. Introduce 10-15 ml of melted media (45° C) in each plate under aseptic conditions.
7. Cover the plates and gently swirl the contents to form thin layer of media with test portion.
8. Allow the medium to solidify and then transfer the plates to incubator. Incubate the plates at 22° C for 4-5 days.
9. Remove the plates and count the colonies formed.
10. Count number of average colonies per plate and multiply with the dilution factor.
11. Compute the count per ml or per g and report.

(ix) Presumptive coliform test

Aim :- To determine presence of coliform bacteria.

Requirements :- Dilution water, Potato dextrose agar, Conical flask, Pipettes, Water bath, Incubator, Petri dishes, test tubes, dilution bottles etc.

Procedure For liquid media

1. Transfer MacConkeys broth in series of market test tubes.
2. Inoculate the tubes with appropriate volumes (10, 1, .01, and 0.001 ml) of sample dilution in triplicate.
3. Incubate the tubes at 37° C for 48 hours.

4. At the end of incubation period check the tubes for any gas formation.
5. Production of acid and gas in two out of three tubes constitute positive test.

For solid media

1. Transfer 1.0 ml of the sample dilution or decimal volumes thereof, into sterile petri plates.
2. Add 10-15 ml of previously melted media (violet red bile agar) and cover the plates.
3. Mix the contents gently and allow to solidify.
4. Distribute additional 3-5 ml of media to completely cover the surface.
5. Incubate the plates at 35° C for 18-24 hours.
6. Presence of dark red colonies constitute positive test.
7. Count dark red colonies having 0.05 mm diameter.
8. Compute count per gram or ml.

Microbiological testing of air

Air is known to be the medium for transmission of organisms like aerobic spore formers, bacteriophages, micrococci, yeast and moulds which may be found in association with dust suspended particles in air. For controlling air contamination it is important to examine number and type of organisms present in the atmosphere inside the dairy plant.

Procedure

1. Pour melted nutrient agar and sabouraud agar in each of two petri plates and allow to set.
2. Remove the top of petri plates and expose each media for 5 minutes in the area of sampling.
3. Incubate nutrient agar plates at 37° C for 48 hours and Sabouroud agar plates at 22° C for 72 hours.
4. Count the number of colonies in each plate.

5. Take a clean and dry glass slide and smear with typical colony in a small area and dry the slide.
6. Dip slides in grams stain for 2 minutes and wash excess of stain in running water.
7. Dry the slides and examine for mould like colonies.
8. Record observation and interpret result as number of organisms per c. ft. per minute.
9. It should not exceed one for satisfactory atmosphere.

Microbiological testing of Water

Natural water from streams, wells and tanks may be having variety of micro-organisms derived from soil, vegetation and other sources. If water is polluted with faecal matter from sewage it is likely to contain pathogenic bacteria. The sanitary quality of water is of importance for dairy industry to avoid contamination. Water is generally examined for total bacterial count and incidence of coliform bacteria.

Sampling

1. Allow the tap to run for 3-4 minutes and stop.
2. Sterilize the tap nozzle by heating with blow lamp or ignited cotton wool soaked in alcohol/spirit.
3. Again allow the tap to run slowly for one minute.
4. Collect sample by holding sample bottle near the tap, removing stopper and quickly placing below the water stream until nearly full. Quickly replace the stopper.

Testing

1. Shake the sample bottle by moving up and down for 25 times.
2. Transfer 1ml of sample directly in duplicate petri plates.
3. Make dilution blanks (1:10 and 1:100) and transfer 1ml solution from each dilution into three set of petri plates in duplicate.

4. Melt prepared tryptone-glucose agar and pour into petri plates, mix evenly and allow to set.
5. Invert plates and incubate one series of plates at 37° C for 48 hours and another series at 22° C for 72 hours.
6. Count the plates having 30-300 colonies.
7. Calculate number of organisms per ml water.

Observations

Sl. No.	Sample source	dilution	Av. Count 37° C	Av count 22° C	Ration of counts	Remark
1						
2						
3						
4						

Note: Samples containing less than 100 bacteria per ml are generally considered to be satisfactory. The ratio of total count at 37° C to that of 22° C should be 1:10. Any change in this ration indicates contamination of water.

Tests for potential pathogenic organisms

The common pathogens of significance in milk are *Escherchia coli Staphylococcus aureus* (coagulase positive), *Salmonella* sp. and *Shigella* sp. The methods of their isolation and identification from milk are complex and time consuming. It is necessary to have a specialized training prior to attempting the screening of samples for these bacteria. The brief protocols for their isolation are given in this text and are based on standard procedures as described in the relevant BIS documents.

A. Isolation and identification of *E.coli*

The typical organism is gram negative, motile, fermenting lactose with the production of gas and usually produces smooth, non-mucoid colonies on solid media. However, there may be strains that are non lactose fermentor. The organism also exhibits following characters; H_2S production, urease negative, indole positive, methyl red positive, growth at 44°C with acid and gas.

Procedure for Enumeration

1. Take two sterile petri plates and transfer to each plate 1ml of test sample (liquid product) or 1ml of initial suspension (other products).
2. Take two other sterile petri plates and transfer to each plate 1ml of first dilution of the test sample or 1ml of first dilution of initial suspension.
3. Repeat procedure with further dilutions.
4. Pour 15ml of VRBL agar medium into each plate, mix the inoculum with media and allow to solidify.
5. After complete solidification, pour 4ml of medium onto the surface and allow to solidify.
6. Prepare a control plate in similar fashion using 15ml medium only.
7. Invert plates and incubate at 37° C for 24 hours.
8. Count the characteristic purple red colonies in each plate containing not more than 150 colonies and not less than 15 colonies.

   Calculation : Calculate number of Coliform per ml or per gram of product using following formula –

$$N = \Sigma c / (Ni + o,1Nii)D$$

Where, N = Number of identified coagulase positive colonies

Σc = Sum of characteristic colonies counted on all plates.

Ni = Number of dishes for first dilution

Nii = Number of dishes for second dilution.

D = Dilution rate corresponding to first dilution (initial suspension)

Procedure for Isolation

- Prepare sample dilutions in 0.1% peptone water.

- Inoculate in MacConkey broth medium, as well as, streak the MacConkey agar and incubate overnight at 37°C.
- Alternatively, pickup characteristic colony from VRBL media and use the same for further identification tests.

Test for identification

- Identify the gram negative rods by staining techniques.
- Test the motility by inoculating in the motility agar and incubation at 37°C for 18-24 hrs.
- The H_2S production is tested on triple sugar iron (TSI) agar medium by streaking the strains into the butt and streaking the slope.
- For urease negative strains, inoculate the urea agar and incubate at 37°C for 18-24 hrs.
- The indole test is carried out by inoculating peptone broth and incubating at 37°C for 48 hours. Add 0.5 ml of Kovac's reagent and observe the appearance of a red colour.
- Strains inoculated in broth are incubated at 37°C for 48 hours. Add 2 drops of methyl red solution to observe for the development of red colour.

B. Isolation and Identification of "*S.aureus*"

The organism is aerobic, gram+ive cocci in clusters, produce shiny black colonies with or without grey-white margin when grown on Baird-Parker medium. Suspect colonies must show coagulase activity. The test is as per IS 5887(Part 8/ section1) :2002.

Media and Reagents

Baird-Parker medium (BPA)

(a) Base- Dissolve by heating in 1000 ml water 10g pancreatic digest casein, 1g Yeast extract, 5g meat extract, 10g sodium pyruvate, 12g L-glycine, 5g lithium chloride, 12-22g agar.

(b) Potassium tellurite solution-Dissolve by heating 1g potassium tellurite in 100 distilled water and sterilize solution using 22 micron membrane filter and store at 3° C. Use before one month.

(c) Egg-yolk emulsion(20%concentration)- Immerse whole eggs in ethanol for 30 seconds and dry. Break and separate yolk from white and place yolk in sterile flask, add 4 times its volume of sterile water, heat in water bath at 47° C for 2 hours and allow to stand for 18-24 hours to form ppt. Aseptically collect supernatant liquid into fresh flask and store at 3° C. Use before 72 hours.

(d)Sulfamezathine-sulfadimidesolution-Dissolve 0.2g sulfamezathine in NaOH(0.1mole/L) and dilute to 100ml with distilled water.Sterilize using membrane filter and store at 3° C. Use before one month.

Preparation of agar plates - To 100ml of base medium add 1ml tellurite solution, 5ml egg-yolk emulsion and 2.5ml Sulfamezathine-sulfadimide solution. Heat all solutions to 47° C before mixing. Pour appropriate quantity of complete medium into petri plates upto 4mm thickness and store plates at 3° C for 24 hours prior to drying. Dry plate with lid off and agar side downwards in oven between 25°-50° C until droplets have disappeared from medium surface.

Brain-heart infusion broth- Dissolve 10g enzyme digest of animal tissue, 12.5g dehydrated calf brain infusion, 5g dehydrated beef heart infusion, 2g glucose, 5g sodium chloride, 2.5g disodium hydrogen phosphate in 1000ml distilled water. Adjust pH 7.4. Transfer culture media @ 5-10ml in tubes or bottles and autoclave at 121° C for 15 minutes.

Rabbit plasma- Dilute 1 volume of fresh rabbit plasma with 3 volumes of sterile water. If sodium citrate is used as anticoagulant, add EDTA solution to give 0.1% EDTA in diluted plasma. Oxalated plasma does not require EDTA. Test each batch of plasma with coagulase positive as well as negative strain of Staphylococci.

Procedure for Isolation

- Prepare serial dilutions from initial suspension.
- Transfer 0.1ml of sample to each of the two agar plates.
- Repeat procedure for further dilutions if necessary.

- Carefully spread inoculum, as quickly as possible, over the surface of agar plate and allow the plates to dry at room temperature for 15 minutes.
- Invert the plates and incubate at 37°C for at least 24-30 hrs.
- Mark on bottom of plate the position of any typical colony. (Typical colonies on BPA are black, shiny with or without a white precipitation and zone of clearance).
- Select five typical and five atypical colonies from each plate for confirmatory test.

Confirmation (Coagulase test)

- From surface of each selected colony, remove an inoculum and transfer to a tube/ bottle containing brain-heart infusion broth.
- Incubate at 37° C for 24 hours.
- Aseptically add 0.1ml of each culture to 0.3ml of rabbit plasma in sterile haemolysis tube and incubate at 37° C.
- Examine the tube for clotting of the plasma after every 4-6 hours of incubation.
- If the volume of clot occupies more than half of original volume of the liquid, the test is considered positive.
- Make a negative control for each batch of plasma by adding 0.1ml of sterile brain-heart infusion broth to recommended quantity of rabbit plasma and incubate without inoculum. This should show no sign of clotting.
- Calculate, for each plate, the number "a" of identified coagulase positive Staphylococci and round off to whole number as follows-

$$A = Bc/Ac \times Cc + Bnc/Anc \times Cnc.$$

Where, A = Number of identified coagulase positive bacteria

Ac = Number of typical colonies submitted to coagulase test.

Anc = Number of atypical colonies submitted to coagulase test

Bc = Number of typical colonies coagulase positive.

Bnc = Number of atypical colonies coagulase positive

Cc = Total number of typical colonies on plate.

Cnc = Total number of atypical colonies on plate

Also calculate number of identified coagulase positive present in test sample as follows-

$$N=\Sigma A/ V(Ni+o,1Nii)D$$

ΣA = Sum of coagulase + ve colonies identified in all selected dishes.

Where, N = Number of identified coagulase positive colonies

V = volume of inoculum (ml) on each dish.

Ni = Number of dishes for first dilution

Nii = Number of dishes for second dilution.

D = Dilution rate corresponding to first dilution (initial suspension)

Present result as-

Number of positive Staphylococci per ml or gm X 10^a

Note: Bacteria of other genera than Staphylococci may give colonies with an appearance similar to Staphylococci. Microscopic examination of gram stained slide, before confirmation, will enable the distinction of other genera from Staphylococci.

C. Isolation and identification of *salmonella* sp.

Salmonellae are gram negative rods, motile, catalase positive, oxidase negative and citrate positive. The serotypes can be differentiated on the basis of Kauffmann-White scheme. While the classification in the subgenera is by biochemical reactions and in the type strains by serological testing.

Media and Reagents:

Rappaport-Vassiliadis(RV) medium-

(solution A) Dissolve by heating, 5g tryptone, 8g sodium chloride, 1.6g potassium dihydrogen phasphate in 1000ml distilled water.

(solution B) Dissolve 400f magnesium chloride hexahydrate in 1000ml distilled water.

(solution C) Dissolve 0.4g malachite green oxalate in 100ml distilled water.

Take 1000ml of solution A, 100ml solution B and 10ml solution C and mix well. Transfer 10ml of this solution into several test tubes and autoclave at 115° C for 15 minutes. Store in refrigerator.

Selenite-cystine medium- (base)- Dissolve by heating, 5g tryptone, 4g lactose, 10g sodium hydrogen phosphate in 1000ml distilled water. Cool the mixture and add 4g sodium hydrogen selenite.

(L-cystine solution)- Dissolve 0.1g L-cystine in 15ml NaOH(1mol/L) and makeup volume to 100ml with sterile water.

Mix 1000ml base with 10ml L-cystine solution and prepare necessary aliquots in sterile flask. Use fresh.

Phenol red- brilliant green agar- (Base)- Dissolve by heating, 5g meat extract powder, 3g yeast extract powder, 10g peptone, 1g disodium hydrogen phosphate, 0.6g sodium dihydrogen phosphate, 12-18g agar in 1000ml water and autoclave at 121° C for 15 minutes.

Brilliant green solution- Dissolve 0.5g brilliant green in 100ml water.

Sugar-phenol red solution- Dissolve 10g lactose, 10g sucrose, 0.09g phenol red in 50ml water and mix well. Makeup the volume to 100 ml with water. Heat at 70° C for 20 minutes and cool. Use fresh.

Mix 100ml sugar-phenol red solution with 1ml brilliant green solution and transfer to 900ml base, heat to 50°-55° C and mix well.

Urea agar- Add 50ml of urea solution(400g urea in 1000ml water, sterilized by membrane filter) to 950ml of base consisting of

1g each of peptone and glucose, 5g sodium chloride, 2g potassium dihydrogen phosphate, 0.012g phenol red, 12-18g agar dissolved in 1000ml water. Dispense mixture in 10ml sterile tubes and allow setting in slanting position.

L-lysine decarboxylation medium- Dissolve by heating, 5g L-lysine monohydrochloride, 3g yeast extract, 1g glucose, 0.015g bromcresol purple in 1000ml water. Transfer in quantities of 5ml in narrow mouth culture tubes and autoclave at 121° C for 10 minutes.

Triple-sugar-iron (TSI) agar - Dissolve by heating, 3g meat extract, 3g yeast extract, 20g peptone, 5g sodium chloride, 10g lactose, 10g sucrose, 1g glucose, 0.3g iron citrate, 0.3g sodium thiosulphate, 0.024g phenol red,12-18g agar in 1000ml water. Dispense in 10ml test tube and autoclave at 121° C for 10 minutes. Allow to set in sloping position to give butt of 2.5 cm depth.

Tryptone-tryptophan medium – Dissolve in 1000ml water at 100° C 10g tryptone, 1g tryptophan, 5g sodium chloride. Dispense in 5ml in several tubes and autoclave at 121° C for 15 minutes.

Kovacs reagent – Mix 5g of 4-dimethylaminobendehyde in 25ml HCl and 2-methylbutan-2-ol.

Nutrient agar – Dissolve by heating 3g meat extract, 5g peptone, 12-14g agar in 1000ml water and autoclave at 121° C for 20 minutes.

Semi-solid nutrient agar - Dissolve by heating 3g meat extract, 5g peptone, 4-9g agar in 1000ml water and autoclave at 121° C for 20 minutes. Place in small petri plates. Do not dry.

Saline solution - Dissolve by heating 8.5g sodium chloride in 1000ml water, dispense in flask to contain 90-100ml and autoclave at 121° C for 20 minutes.

Procedure

1. To 225ml of pre-enrichment medium(buffered peptone water) add 25g of test material and incubate the suspension at 37° C for 16 hours. (Non-selective enrichment)
2. Transfer 0.1ml of the culture obtained in a test tube containing 10ml RV medium and incubate at 42° C for 18-24 hours. (Selective enrichment)

3. Transfer 10ml of the culture obtained in a flask containing 100ml selenite-cystine medium and incubate at 37° C for 18-24 hours.
4. Inoculate a loop full of cultures of RV medium culture in large size petri dish containing phenol red- brilliant green agar.
5. Proceed in same way with second selective plating media.
6. Repeat procedure for inoculating culture from selenite-cystine medium with two selective plating media.
7. Invert the plates and place in incubator at 37° C for 24 hours.
8. Examine the plate for presence of typical colonies that change the media colour from pink to red.
9. Re-incubate in absence of typical colonies for another 24 hours.

Note : Use pure cultures as control for biochemical and serological confirmation.

Confirmation tests

1. Take from each plate of selective medium five typical colonies.
2. Streak selected colonies onto dried nutrient agar plates for development of isolated colonies.
3. Incubate plates at 37° C for 18-24 hours.
4. Use culture/colonies for further biochemical and serological confirmation.

Biochemical confirmation

TSI agar test: Inoculate the TSI media (streak slops and stab the butts) with each of the culture obtained from selected colonies. Incubate at 37° C for 24 hours. Typical Salmonella culture results in blackening of the agar due to formation of hydrogen sulphide.

Indol reaction – Take 5ml of tryptone-tryptophan medium in a test tube, inoculate with suspected colony and incubate at 37 C for 24 hours. Add 1ml of Kovacs reagent. Formation of red ring indicates positive reaction. While yellow colour indicates negative reaction.

Urea agar test– Streak urea agar slop with suspected colony, incubate at 37° C for 24 hours. Examine at interval of 2-4 hours for change of colour of phenol red to pink to deep cerise, indicating positive reaction.

L-Lysine decarboxylation medium test – Inoculate just below medium surface with suspected colony and incubate at 37° C for 24 hours. Purple colour indicates positive reaction. While yellow colour indicates negative reaction.

Serological confirmation: Test by slide agglutination with appropriate sera in presence of antigen, after elimination of auto-agglutinating strains.

Elimination of auto-agglutinating strains: Place one drop of saline solution on a clean and dry glass slide and disperse part of colony to obtain turbid suspension. Dry the slide and observe under magnification, any clumping of bacteria into distinct units. If clumping exist eliminate strain. If no clumping appears use strain for further antigen tests.

"O" antigen test – Select non-auto agglutinating colony. Place a drop of anto-O serum on the slide and dispense colony into the serum. Dry the slide and examine. If agglutination occurs, the reaction is positive. Use poly and monovalent sera one after another.

"Vi"-antigen test – Use one drop of anti Vi serum on the slide and dispense colony into the serum. Dry the slide and examine. If agglutination occurs, the reaction is positive.

Interpretation for changes in media

Slant surface	**Butt**
Yellow —— lactose / sucrose positive	Yellow — Glucose positive
Unchanged/Red– lactose/sucrose negative	Red —— Glucose negative
	Black – formation of Hydrogen sulphide
	Bubbles/cracks – Gas formation

Interpretation of confirmatory tests

Biochemical reaction	Auto-agglutination	Serological reaction	Interpretation
typical	No	Antigen positive	Strain considered to be Salmonella
Typical	No	All reaction -ve	May be Salmonella
Typical	yes	Not tested	
No typical reaction	No	Antigen positive	
No typical reaction	No	All reaction -ve	Not considered Salmonella

Strains considered or may be considered shall be sent to a recognized Salmonella reference center for definitive typing.

Result : Indicate presence or absence of Salmonella in a test portion of xg of product, based on interpretation.

D. Isolation and identification of "*Shigellae*" spp.

Shigellae are also gram-negative rods, non-motile, catalase positive, oxidase negative and citrate negative. The genus has been divided on the basis of biochemical characteristics and serological testing.

Media and reagents

Nutrient broth – Dissolve by heating 10g peptone, 10g meat extract and 5g sodium chloride in 1000ml water. Filter and remove ppt. Autoclave at 121° C for 15 minutes.

Nutrient agar – Add 1.5-3% agar to the above ingredients.

Selenite F broth – Dissolve by heating 4g sodium acid selenite, 5g peptone, 4g lactose, 9.5g disodium hydrogen phasphate, 0.5g sodium dihydrogen phasphate in 1000ml water. Transfer 10ml aliquots in screw cap bottles and steam at 100° C for 30 minutes.

Deoxycholate citrate agar

1. **Base** – Mix 20g meat extract in 200ml water and make it

alkaline with 50% NAOH solution. Add 20f proteose peptone. To another vessel dissolve 90g agar in 3700ml water by steaming for one hour. Filter and add to it meat extract-peptone solution and mix. Add 5ml of 2% neutral red solution in 50% ethanol and 40g lactose. Make upto 4L with water and sterilize at 120° C for 10 minutes.

2. **Solution A** – Dissolve by heating 17g sodium citrate, 17g sodium thiosulphate and 2g ferric ammonium citrate in 100ml sterile water.

3. **Solution B** – Dissolve 10g sodium deoxycholate in 100 ml sterile water.

Sterile solution A and B at 60° C for one hour. Melt 100ml agar base and add 5ml of each solution A and B. Mix well and pour 12-15ml into sterile petri plates and allow to set. Dry surface in incubator before use.

MacConkeys agar – Mix 5g bile salts (sodium tauocholate), 20g peptone, 5g sodium chloride and 15-30g agar in 1000ml water by heating, filter and add 10g lactose and 3.5ml of 2% solution of neutral red in 50% alcohol. Mix and distribute in flask and autoclave at 121° C for 15 minutes. The consistency of the agar should be soft. Pour into petri dishes and allow to set.

Hugh-Leifsons medium – Dissolve by heating 2g peptone, 5g sodium chloride, 0.3g disodium hydrogen phasphate and 3g agar in 1000ml water. Filter and add 15ml of 0.2% alcoholic solution of bromthymol blue. Autoclave at 115° C for 15 minutes.

Kauffmann- Muller tetrathionate broth – To 90ml sterile nutrient broth add 5g of powdered and sterilized calcium carbonate, 10ml 0f 50% solution of sodium thiosulphate, 2ml potassium iodine solution(grind 5g potassium iodide in 20ml water and 4g iodine), 1ml of 0,1% brilliant green and 1g bile salt(ox bile).

TSI medium – Dissolve by heating 20g peptone, 3g meat extract, 3g yeast extract, 1g glucose, 10g lactose, 10g sucrose, 0.2g ferrous sulphste, 5g sodium chloride, 0.3g sodium thiosulphate and 15-30g agar in 1000ml water. Add 12ml of 0.2% phenol red solution. Pour into test tubes and autoclave at 115° C for 20 minutes. Cool by positioning tubes in slate to form slops with deep butts.

Urease medium- Dissolve by heating 1.5g peptone, 5g sodium chloride, 2g dipottasium hydrogen phosphate and 15-30g agar in 1000ml water. Add 6ml of 1:500 aqueous solution of phenol red and autoclave at 120° C for 15 minutes. Cool and add sterile solution of glucose to give final concentration of 0.1% and 100ml of 20% sterile solution of urea. Distribute in sterile test tubes to form deep slops.

Simmons citrate agar – Dissolve 5g sodium chloride, 0.2g magnesium sulphate, 1g ammonium dihydrogen phosphate, 1g dipottasium hydrogen phosphate, 2g sodium citrate and 15-30g agar; add 40ml of 1:500 bromthymol blue indicator solution and autoclave at 120° C for 15 minutes. Pour in sterile tubes to form equal butts and slops.

Indole medium – Dissolve 20g peptone and 5g sodium chloride in 1000ml water. Place in tubes 5ml portions and autoclave at 120° C for 15 minutes.

Peptone–water medium – Dissolve 10g peptone and 5g sodium chloride in 900ml water, add 10ml of Andrades indicator(add1N NaOH in 0.5% aqueous solution of acid fuchsin until indicator colour is just yellow) and 90ml of sterile sugar solution(10g sugar +90ml water). Distribute into sterile test tubes containing Durham's tubes and steam for 30 minutes

Isolation

- Pre-enrich, in nutrient broth, all food that has received dry heat treatment, dehydration and desiccation.
- Raw food and finished food suspected of post contamination does not require pre enrichment.
- Blend 25g sample in sterile jar for 2 minutes using 200ml selenite F broth and incubate at 37°C for 24 hours.
- Blend another portion of 25g sample in 200ml tetrathinate broth and incubate at 37°C for 24 hours.

- From each incubated sample plate a loop full on Desoxycholate citrate agar medium (DCA).
- Also plate out loop full of each of the liquid medium (blended sample) onto solid media and incubate plates at 37° C for 24 hours.
- Examine for suspect colonies. These colonies on DCA are opaque with ground glass appearance and with even margins.
- Reincubate for another 24 hours if no suspected colonies are observed.
- Pick out as many suspected colonies as possible for further investigation.
- Plate out onto McConkeys agar and incubate at 37° C overnight.
- Transfer pure colonies onto nutrient agar to obtain heavy growth for preserving isolates for identification.

Tests for Identification

1. **Gram stain test** – Take a clean grease free slide and smear a thin layer of culture in a small area. Fix by passing to and fro over a flame and cool. Cover with stain A (0.5% methyl violet/crystal violet in water) for 30 seconds. Pour off stain, wash and cover with stain B (1%iodine + 2% potassium iodide in water) for 30 seconds. Wash slide with ethanol and remove excess stain by washing in tap water. Apply stain C (0.1g neutral red, 0.2ml of 1% acetic acid and 100ml water) for 1 minute, wash in tap water , dry and examine.
2. **Catalase test** – Grow strain on nutrient agar slop for 24 hours. Pour 1ml hydrogen peroxide over the growth with tube in slanting position. Release of oxygen shown as bubbles indicate presence of catalase.
3. **Oxidase test** – Add few drops of testing reagent (1% solution

of alpha-naphthol in 95% ethanol + equal volume of 1% solution of para-amino-dimethyl analine hydro chloride in water) to the culture grown in nutrient agar slant. Appearance of blue colour within two minutes indicate positive reaction.

4. **Urase test** – Take loop full of culture from nutrient broth after 24 hours incubation, inoculate over entire slop of the medium and incubate at 37° C for 18-24 hours. Observe change in colour, of medium. A pink or red colour indicate positive urase. If negative continue incubation for at least 4 hours.

5. **Indole test** – Take loop full of culture from nutrient broth after 24 hours incubation and inoculate on indole production medium. Incubate at 37° C for 48 hours and add 0.5ml Kovacs reagent and shake tube gently. Appearance of red colour indicates presence of indole.

6. **Hugh-Leifsons test** – Grow culture on nutrient agar. Take a loop full and stab into two tubes containing Hugh-Leifsons medium. Layer one tube with sterile liquid paraffin. Incubate tubes at 37° C for 4 days. Observe tubes every 24 hours. Formation of yellow colour and acid in tube No 2 (without paraffin) indicates oxidative utilization of glucose. Acid formation in both tubes indicates fermentative reaction. Lack of acid indicates strain not able to utilize glucose.

7. **Carbohydrate fermentation test** – Inoculate each of peptone water medium and carbohydrate indicated in table 1 using 1% concentration of lactose. Incubate at 37° C for 18-24 hours. Record presence of acid from pink colour and that of gas in Durhams tube.

8. **H_2S production** – Inoculate TSI medium by stabbing the strain into the butt and streaking the slop. Incubate at 37° C for seven days. Daily observe for presence/ absence of blackening in the butt of the medium and record.

Salmonella **and** ***Shigella*** are suspect when conforming to the characteristics as given in Table. 1

Some useful characteristics of Salmonella and Shigella

S.No.	Tests	*Salmonella* Subgenus I	*Shigella* Subgroup A
1	Gram negative rod	+	+
2	Motility	+	-
3	Catalase	+	+
4	Oxidase	-	-
5	TSI for H2S	+	-
6	Urease	-	-
7	Citrate	+	-
8	Indole	-	Variable
9	Phenyl pyruvic acid	-	-
10	Glucose	+	+
11	Lactose (1%)	-	-
12	Salicin	-	-
13	Dulcitol	+	-
14	Gelatine	-	-
15	Malonate	-	-

E. Isolation and identification of "*Listeria monocytogenes*"

Preparation of media and reagents

Buffered peptone water - Dissolve 10 g enzymatic digest of animal tissues, 5g Sodium chloride (NaCl), 9g Disodium hydrogen phosphate dodecahydrate ($Na_2HP_041\ 2H2_0$), 1.5g Potassium dihydrogen phosphate ($KH2P_04$) in 1000ml water, by heating if necessary. Adjust the pH [measured with the pH-meter (6.7)], if necessary, so that after sterilization it is 7.0 +/- 0.2 at 25^0C. Dispense the medium in flasks of suitable capacity to obtain portions appropriate for the test. Sterilize for 15 min in the autoclave set at 121^0C.

Half-Fraser broth with ammonium iron (III) citrate

Base medium - Dissolve 5g Enzymatic digest of animal

tissues, 5g Enzymatic digest of casein, 5g Meat extract, 5g Yeast extract, 20g Sodium chloride (NaCl), 12g Disodium hydrogen phosphate dehydrate (Na2HP04.2H20), 1.35 g Potassium dehydrogenate phosphate (KH2P04) and 1g Aesculin in 1000ml water by heating if necessary. Adjust the pH, if necessary, so that after sterilization it is 7.2 +/- 0.2 at 25°C. Dispense the base in flasks of suitable capacity to obtain portions appropriate for the test. Sterilize for 15 min in the autoclave set at 121°C.

Ammonium iron (lll) citrate solution - Dissolve 5g ammonium iron (lll) citrate in 100ml water. Sterilize by filtration.

Complete medium - Mix 1ml of ammonium iron citrate solution to 100ml portion of base medium, immediately before use.

Selective plating-out medium: PALCAM agars

Agar base - Dissolve 23g Peptones, 1g Starch, 5g Sodium chloride (NaCl), 3g Yeast extract, 9 to 18g Agar, 0.5g D-Glucose, 10g D-Mannitol, 0.8g Aesculin, 0.5g Ammonium iron (lll) citrate, 0.08g Phenol red and 15g Lithium chloride (LiCl) in 960ml water, by boiling. Adjust the pH, if necessary, so that after sterilization it is7.2 +/- 0.2 at 25°C. Sterilize for 15 min in the autoclave set at 121°C.

Polymyxin B sulfate solution - Dissolve 0.1g polymyxin B sulfate (100,000 IU) in 100ml water. Sterilize by filtration.

Acriflavine hydrochloride solution - Dissolve 0.05g acriflavine hydrochloride in 100ml water. Sterilize by filtration.

Sodium ceftazidime pentahydrate solution - Dissolve 0.116g sodium ceftazidime in 100ml water. Sterilize by filtration.

Complete medium - Add 10 ml Polymyxin B sulfate solution, 10ml Acriflavine hydrochloride solution and 20ml Sodium ceftazidime dentahvdrate solution to 960ml molten base at 47°C, mixing thoroughly between each addition. The pH of the complete medium shall be 7.2 +/- 0.2 at 25°C.

Tryptone Soya Yeast Extract Agar (TSYEA) - Dissolve 17g Enzymatic digest of casein, 3g Enzymatic digest of soya meal, 5g Sodium chloride (NaCl), 2.5g Dipotassium hydrogen phosphate (K2HPO4), 2.5g D-Glucose, 6g Yeast extract, 9 to 18g Agar in

1000ml water by boiling. Adjust the pH, if necessary, so that after sterilization it is 7.3+/- 0.2 at 25°C. Dispense the medium into tubes of suitable capacity to obtain portions appropriate for the test. Sterilize for 15 min in the autoclave set at 121°C. Allow to set in a sloping position.

For the preparation of agar plates, dispense the medium into sterile Petri dishes in portions appropriate for the test. Allow to solidify.

Tryptone Soya Yeast Extract Broth (TSYEB) - Dissolve 17g Enzymatic digest of casein, 3g Enzymatic digest of soya meal, 5g Sodium chloride (NaCl), 2.5g Dipotassium hydrogen phosphate (K2HPO4), 2.5g D-Glucose and 6g Yeast extract in 1000ml water, by heating if necessary. Adjust the pH, if necessary, so that after sterilization it is 7.3 +/- 0.2 at 25°C. Dispense the medium into flasks or tubes of suitable capacity to obtain portions appropriate for the test. Sterilize for 15 min in the autoclave set at 121°C.

Sheep blood agar

Base - Dissolve 15g Enzymatic digest of animal tissues, 2.5g enzymatic liver digest, 5g Yeast extract, 5g Sodium chloride (NaCl), and 9 to 18g Agar in 1000 ml water by boiling. Adjust the pH, if necessary, so that after sterilization it is 7.2 +/- 0.2 at 25 ^{0}C. Dispense the medium into flasks of suitable capacity to obtain portions appropriate for the test. Sterilize for 15 min in the autoclave set at 121 ^{0}C.

Complete base - Add 5 to 7ml sheep blood to 100ml of the base previously cooled to about 47 ^{0}C. Mix well. Dispense the medium into sterile Petri dishes in portions appropriate for the test. Allow to solidify.

Carbohydrate utilization broth (rhamnose and xylose)

Base - Dissolve 10g Enzymatic digest of animal tissues, 1g Meat extract, 5g Sodium chloride (NaCl) and 0.02g Bromocresol purple in 1000ml water, by heating if necessary. Adjust the pH, if necessary, so that after sterilization the pH is 6.8 +/- 0.2 at 25 ^{0}C.

Dispense the medium into tubes of suitable capacity to obtain portions appropriate for the test. Sterilize for 15 min in the autoclave set at 121°C.

Carbohydrate solutions - Dissolve separately 5g carbohydrate in 100 ml of water. Sterilize by filtration.

Complete media - For each carbohydrate, add aseptically x ml of carbohydrate solution to 9X ml of the base

Motility agar - Dissolve 20g Enzymatic digest of casein, 6.1g Enzymatic digest of animal tissues and 3g Agar in1000ml water by boiling. Adjust the pH, if necessary, so that after sterilization it is 7.3 +/- 0.2 at 25°C. Dispense the medium into tubes in quantities of about 5 ml. Sterilize for 15 min in the autoclave set at 121°C.

CAMP (Christie, Atkins, Munch-Paterson) medium and test strains

Sheep blood agar plates maybe used for this test, but it is preferable to use double-layered agar plates with a very thin layer of sheep blood medium

Base - Dissolve 15g Enzymatic digest of animal tissues, 2.5g enzymatic liver digest, 5g Yeast extract, 5g Sodium chloride (NaCl), and 9 to 18g Agar in 1000 ml water by boiling. Adjust the pH, if necessary, so that after sterilization it is 7.2 +/- 0.2 at 25°C. Dispense the medium into flasks of suitable capacity to obtain portions appropriate for the test. Sterilize for 15 min in the autoclave set at 121°C.

Sheep blood medium - Add 5 to 7ml sheep blood to 100ml of the base previously cooled to about 47 ^{0}C. Mix well. Dispense the medium into sterile Petri dishes in portions appropriate for the test. Allow to solidify.

Complete medium - Dispense the base into sterile Petri dishes, in quantities of about 12 ml for Petri dishes of 90 mm of diameter, and allow to solidify. Pour evenly a very thin layer of the sheep blood medium using amounts not greater than 3 ml per plate. Allow to solidify. If the blood medium is added to dishes containing the base, which have been prepared in advance, it may be necessary to warm the dishes for 20 min by placing them in an incubator set at 37 ^{0}C before pouring the blood medium layer.

CAMP reaction strains

To undertake the CAMP test, maintain stock cultures of a ?-haemolytic strain of S. *aureus (e.g.* NCTC 1803 or ATCC 25923), *R. equi (e.g.* NCTC 1621 or ATCC 6939), *L. monocytogenes (e.g.* NCTC 11994), *L. innooua* and *L. ivanovii* by inoculating the TSYEA slopes, incubating them at 37 ^{0}C for 24 h to 28 h, or until growth has occurred, and storing in the refrigerator at +3 ^{0}C. Subculture at least once per month, and confirm their purity by streaking on TSYEA plates. Not all strains of S. *aureus* and L. *monocytogenes* are suitable for the CAMP test.

Hydrogen peroxide solution - Use a **3 % (m/m),** i.e. 10 volume, solution.

Phosphate-buffered saline (PBS) - Dissolve 8.98g Disodium hydrogen phosphate dihydrate (Na2HPO4.2H2O), 2.71g Sodium dihydrogen phosphate (Na2HP04) and 8.5g Sodium chloride (NaCl) in 1000 ml water. Adjust the pH, if necessary, so that after sterilization it is 7.2 +/- 0.2 at 25°C.Sterilize in the autoclave for 15 min at 121°C.

Sheep blood red cell suspension (optional) - Keep the sheep blood cells at +3°C +/- 2°C before use. Examine before use for signs of haemolysis (reddening) in the upper serum layer. If no haemolysis is present, take *2* ml of the lower red cell layer into 98 ml of PBS buffer. If some haemolysis has occurred, suspend about 4 ml of the red cell layer in 10 ml of PBS buffer and mix gently, then centrifuge. If the supernatant is obviously reddened by significant haemolysis, do not use and discard the stock suspension. Otherwise decant the supernatant and add 2 ml of this cells suspension to 98 ml of PBS buffer. Store the suspension for up to 5 days at +3°C + 2°C. Discard if any haemolysis occurs.

Sample preparation : Mix 1g of sample to 9ml of buffered peptone water or Half Fraser broth to get initial suspension. Also make series of dilutions from initial suspension for further enumeration and confirmatory tests.

Determination

1. Transfer 0.1ml of the initial suspension to each of two petri plates containing 10-12 ml of PALCAM agar.

2. Repeat procedure using further decimal dilutions.
3. Carefully spread inoculum over the surface of agar using fresh sterile spreader for each plate.
4. Close the petri plates and leave for 15 minutes at room temperature for inoculum to absorb into the agar.
5. Invert plates and incubate at 37°C for 24 hours.
6. If growth is slight or no colony is observed, re-incubate for another 18-24 hours at 37°C.
7. Examine plates for presence of characteristics colonies of Listeria spp. Which grow as small (1.5-2.0 mm dia) grayish green or olive green with black halos.
8. Count colonies on each plate containing less than 150 colonies.

Confirmation

9. Select five colonies in each plate.
10. Streak selected colonies onto surface of TSYEA agar in a manner to allow well separated colonies to develop.
11. Incubate at 37 ^{0}C for 24 hours.
12. Typical colonies are convex, colourless, and opaque with entire edge.
13. Pick a typical colony onto another TSYEA plate and carry out further tests.

Gram staining – Smear, from a colony, on a small portion of clean and dry slide. Dip in Grams stain for 2 minutes and wash excess of stain. Examine under microscope for gram positive , slim short rods.

Catalase reaction – Using inoculating needle suspend a typical colony on a drop of hydrogen peroxide solution on a slide. Immediate formation of gas bubbles indicate positive reaction.

Alternative Motility test – Stab the motility agar with a typical colony on TSYEA. Incubate at 25 ^{0}C for 48 hours. Examine for typical umbrella like growth pattern, below the agar surface. If not sufficient incubate for another five days and examine during this period.

CAMP Test

1. Disperse the colony in 150 µl of TSYEA and incubate at 37°C for 2 hours. Add 150 µl of sheep blood red cell suspension and incubate for 15-60 minutes and then refrigerate for 2 hours. Examine for presence or absence of ß-haemolysis.
2. Streak each of *S. aureus* and *R. equi* cultures in single line across sheep blood agar plate, so that the two cultures are parellel. Streak the test strain at right angle to these cultures so that they do not touch but are 1-2 mm apart. Similarly, streak control culture of Listeria spp. Incubate the plates at 37°C for 24 hours. An enhanced zone of ß-haemolysis at the intersection of test strain with cultures of *S. aureus* and *R. equi* is considered positive.

Haemolysis test

1. Pour sheep blood agar into petri plates and allow to dry. Mark agar with squares.
2. Take a colony, for each culture, and stab one labeled square using inoculating wire.
3. Also stab positive and negative control cultures.
4. Incubate at 37 ^{0}C for 24-26 hours and examine both, test strains and control.
5. *L. monocytogenes* shows narrow, clear, light zone of ß-haemolysis.
6. *L. innooua* shows no clear zone around stab.
7. *L. ivanovii* shows wide, clearly delineated zones of ß-haemolysis.
8. *L. seeligeri* shows weak zones of ß-haemolysis.

Reaction for identification of Listeria spp.

Species	Haemolysis	Acid production		CAMP Test	
		Rhamnose	Xylose	*S . aureus*	*R. equi*
L. monocytogenes	+	+	-	+	-
L. innooua	-	V	-	-	-
L. ivanovii	+	-	+	-	+
L. seeligeri	(+)	-	+	(+)	-
L.weishimeri	-	V	+	-	-
L.grayi	-	-	-	-	-

V: Variable reaction; (+) =:Weak reaction; + : Positive reaction (90%); - : No reaction

Note : Strains showing positive test should be sent to Listeria reference laboratory for serological, lysogenic and molecular typing methods.

Colony counting : Calculate for each plate the number of colonies using formula

$$A = B / C \times D$$

A = Number of typical colonies from all plates.

B = Number of colonies confirming to the identification criteria.

C = Number of colonies plated out for confirmation.

D = Total number of characteristic colonies enumerated on the plate

Calculation: calculate number of identified *L.monocytogenes* present in test sample as follows-

$$N= \Sigma A/ V(Ni+o,1Nii)D$$

ΣA = Sum of the confirmed colonies on all dishes

Where, N = Number of identified colonies

V = volume of inoculum (ml) on each dish.

Ni = Number of dishes for first dilution

Nii = Number of dishes for second dilution.

D = Dilution rate corresponding to first dilution (initial suspension)

Present result as

Number of *L. monocytogenes* per ml or gm X 10

❑❑❑

Chapter–4

Determination of Minerals and Vitamins in Milk

A. Estimation of Calcium

Requirements : Pottasium permanganate solution (0.1N) – Dissolve by heating 33g of potassium permanganate in 1000ml distilled water and allow to stand in coloured bottle for several days. Filter contents through 6mm thick film of asbestos mat on Gooch crucible.

Asbestos mat – Digest asbestos with dilute hydrochloric acid(1:3) for 2-3 days, wash free from acid and digest with 10% sodium hydroxide for few days, wash free from alkali. Digest with dilute Nitric acid(1:3) for several hours and wash. Shake with water into fine pulp and digest with 0.1N potassium permanganate solution that has been acidified with dilute sulphuric acid (1:3), for 1 hour and wash. Use this pulp for layering 6mm film on Gooch crucible.

Dilute hydrochloric acid (1:4), Ammonium hydroxide solution (1:1), Saturated ammonium oxalate solution, Diluted sulphuric acid (1:4)

Procedure 1 (Titration Method)

1. Dissolve ash, obtained from determination of ash in milk sample, in dilute hydrochloric acid (1:4), transfer to 100ml flask and makeup the volume.
2. Transfer 20ml of diluted ash solution to a beaker and add 30ml water.
3. Add excess of ammonium hydroxide to make slightly alkaline.
4. Add dilute ammonium hydroxide solution (1:1) drop wise, heat to boiling and allow the precipitate to form. Allow to stand for 3 hours.
5. Decant clear solution through Whatman No. 42 filter paper and wash precipitate with 15-20ml hot water.
6. Dissolve any precipitate remaining on filter paper by washing with hot diluted hydrochloric acid (1:9), into original beaker and wash 6 times with hot water.
7. Re-precipitate by addition of ammonium hydroxide and little saturated ammonium oxalate solution and allow to stand.
8. Filter through same filter paper and wash with hot water until chloride free.
9. Collect filtrate and washings and use for determination of magnesium.
10. Perforate apex of filter paper cone and wash the precipitate into the beaker
11. Wash filter paper with hot diluted sulphuric acid (1:4) and titrate with 0.1N potassium permanganate solution.

    Calculation : Calcium, percent by weight = $1.002 \times V / W$

    Where, V = ml of 0.1N potassium permanganate used for titration.

    W = Weight in g of prepared sample taken for ash estimation.

Procedure 2 (Flame emission spectrometric Method)

Preparation of standard solution – Weigh 2.497g of calcium carbonate, transfer to 1000 ml volumetric flask and make up the volume with distilled water.

Preparation of reference solution - Pipette 5, 10, 15, 20, and 25 ml this solution in different 1000 ml volumetric flasks and add 700 ml water and mix. Add to each flask 10 ml of EDTA solution (25g in 1000 ml), neutralize with sodium hydroxide (0.4 mol/L) solution and make up the volume with water and mix well.

This solution corresponds to 5,10,15,20 &25 mg of Ca per liter.

Preparation of test solution – Take 1.0g of test material in 100 ml volumetric flask and dissolve in 20 ml of hot water (50 C). Cool and make up the volume with distilled water. Pipette 25 ml of the solution in 100 ml volumetric flask and make up the volume with distilled water. Pipette 10 ml of the solution in 100 ml volumetric flask; add 70 ml water, 0.2 ml of NaOH solution and 1 ml EDTA solution. Mix by swirling after each addition. Make up the volume and mix well.

Determination – Start the flame spectrometer by supplying mixture of acetylene and air to the burner. Insert the calcium filter with maximum transmittance at 623 nm. Inject reference solution starting with lowest concentration and the test solution alternatively into the flame of spectrometer using calcium filter and note the reading.

Calculation :- calcium content (%) = { (Ts – Lr / Hr – Lr) x (Q2 – Q1) + Q1} f/m

Where -

Ts	=	reading of test solution
Lr	=	Lowest reading for reference solution of conc. Q1
Hr	=	Highest reading of reference solution of conc. Q2
Q1	=	Concentration of reference solution (mg/ L) giving reading Lr
Q2	=	Concentration of reference solution (mg/ L) giving reading Hr
f	=	Conversion factor for calcium (0.1)
m	=	weight of test sample (g)

B. Estimation of Magnesium

Requirements : Phosphate solution – Dissolve 100g ammonium hydrogen phosphate in 1000ml water by heating and add 5 drops of chloroform to preserve.

Citric acid solution (1M), Alcohol (95%), Dilute ammonium hydroxide (1:9).

Ammonium hydroxide (sp.gr.0.88)., Ammonium nitrate solution (saturated ammonical solution).

Procedure

1. To the combined filtrate obtained from calcium estimation, add 2ml of 1M citric acid solution, 100ml ammonium hydroxide and 50ml of 95% alcohol.
2. To this mixture add 25ml of phosphate solution with constant stirring.
3. Allow to stand for 12-20 hours. Filter through Whatman No 42 paper and wash twice with dilute ammonium hydroxide (1:9).
4. Dissolve precipitate in dilute hydrochloric acid (1:4) in a graduated beaker to make volume of 100-150ml.
5. Add 10-15ml ammonium hydroxide and 2 drops of phosphate solution, stir and allow to stand for 3 hours.
6. Filter through Gooch filter and wash with dilute ammonium hydroxide (1:10).
7. Moisten filter with dilute ammonium nitrate solution (saturated ammonical solution).
8. Ignite and weigh as magnesium pyro-phosphate.

Calculation : Magnesium, percent by weight = $109.24 \times A/W$

Where, A = mg of magnesium pyro-phosphate

W = Weight in g of prepared sample taken for ash estimation.

C. Estimation of Phosphorus

Requirements : Molybdate solution- Dissolve 100g molybdenum oxide in a mixture of 144ml ammonium hydroxide and 271ml water by heating. Cool and add slowly into it the cold mixture of 489ml nitric acid and 1148ml water. Store for several days at room temperature, filter and store in coloured glass stoppered bottle. Take 100ml of this solution and add 5ml nitric acid, filter immediately before use.

Sodium / Potassium hydroxide – Dilute 324ml of 1N sodium / potassium hydroxide to 1L with water, Ammonium hydroxide, 2% sodium nitrate, 0.1N hydrochloric acid or Nitric acid.

Procedure 1

1. Take 20ml of diluted ash solution and add 10ml of nitric acid.
2. Add ammonium hydroxide until precipitate formed is dissolved.
3. Dilute to 75-100ml and add 20-25ml of freshly filtered molybdate solution.
4. Stirr contents for 30 minutes and filter through Whatman No.42 filter paper.
5. Wash precipitate twice with 20-30ml water containing 2% sodium nitrate.
6. Transfer precipitate and filter paper to a beaker.
7. Dissolve precipitate in little excess of sodium/potassium hydroxide solution and note volume of alkali used.
8. Add few drops of indicator and titrate with standard acid solution.
9. One ml of standard alkali solution = 0.4364 mg phosphorus.

   Calculation: Phosphorus, percent by weight = 0.4364 (A-B) / 2W

Where, A= ml of standard alkali used for dissolving precipitate.

B = ml of standard acid used for titration.

W = Weight in g of prepared sample taken for ash estimation.

Procedure 2

Requirements :- Volumetric flasks (50,100,1000 ml), spectrophotometer, beaker, weighing balance, water bath, sulphuric acid (5mol/L), sodium molybdate solution (2.5g dissolved in sulphuric acid and diluted to 100 ml with acid), Ascorbic acid (5g diluted to 100 ml with water), molybdate-ascorbic acid solution(mix 25 ml of sodium molybdate solution and 10 ml of Ascorbic acid solution and diluted to 100 ml with water.

Preparation of standard solution – Weigh 0.439g of dried potassium hydrogen orthophosphate, transfer to 1000 ml volumetric flask and make up the volume with distilled water. (equivalent to 100 ug of P/ml)

Preparation of diluted solution – Transfer 10 ml of standard solution into100 ml volumetric flask and make up the volume with distilled water and mix well. (equivalent to 10 ug of P/ml).

Preparation of calibration curve – Pipette 0,1,2,3 and 5 ml of diluted solution in series of 50 ml volumetric flasks. Add 25 ml water and 2 ml of molybdate- ascorbic acid solution. Dilute to mark and mix well. Place the flask in boiling water bath for 15 minutes. Cool and measure absorbance at 820 nm against the reagent blank. Plot the absorbance values against the amount of phosphorous added.

Blank test – Follow same procedure as for sample but with 1, 5 or 10 ml of phosphorous free water instead of test sample.

Sample preparation – Take the ash, obtained after ashing, and dissolve in 2-3 ml of hydrochloric acid. Transfer the solution to 100 ml volumetric flask and make up the volume. Filter the contents and collect filtrate. Pipette 10 of filtrate into 100 ml volumetric flask, make up the volume and mix well.

Determination – Pipette 2 ml of test solution into 50 ml volumetric flask. Add 25 ml water and 2 ml of molybdate ascorbic acid solution. Dilute to mark and mix well. Place the flask in boiling water bath for 15 minutes. Cool and measure absorbance at 820 nm against the reagent blank. Plot the result in calibration curve to know the concentration.

Calculation :- Total phosphorous (%) = M1 / 20 m

Where- M1 = Phosphorous concentration as read from calibration curve.

m = weight of test sample (g)

D. Estimation of Iron

Requirements : Pottasium persulphate solution (2%) made in double distilled water and stored under refrigeration. Prepare fresh before use.

Pottasium thiocynate solution (20%) made in double distilled water.

Bromine water (saturated bromine solution), Anhydrous sodium sulphate, Hydrochloric acid (20%), iso-butyl-alcohol, Dilute hydrochloric acid (1:1)

Preparation of standard curve

Stock solution - Weigh 1.0g of iron wire in a clean and dry beaker and dissolve in sufficient hydrochloric acid to which 1-2 drops of nitric acid is added. Evaporate to dryness and dissolve residue in concentrated hydrochloric acid. Transfer to 100ml volumetric flask and dilute to volume.

Working solution

1. Prepare a working standard of 10 μg/ml by diluting with water. Add few drops of bromine water prior to adjusting volume.
2. Take required volume of standard iron solution, to give a range of 0-60 μg of iron, in series of 25 ml volumetric flask.
3. Add 5ml hydrochloric acid, 1ml potassium persulphate and 10ml potassium thiocynate solution and make up the volume.
4. Transfer small quantity to colorimeter cuvette and observe transmission at 485 nm.
5. Draw a standard curve between transmission and concentration.

Procedure

1. Transfer 3-5g of sample in silica crucible, char at low heat and place in muffle furnace at 550° C for 5 hours.

2. Dissolve ash in 5ml dilute hydrochloric acid (1:1) with gentle boiling for 3 minutes.
3. Cool and transfer to 100 ml flask and make up the volume with distilled water.
4. Transfer 25ml of prepared solution to 125 ml separating funnel.
5. Add 5ml concentrated hydrochloric acid, 1ml potassium persulphate and 10ml potassium thiocynate solutions.
6. Mix well and add 25 ml of iso-butyl-alcohol and shake for 2 minutes.
7. Discard the aqueous layer and transfer alcohol layer to 25 ml measuring cylinder.
8. Add 1.0 g anhydrous sodium sulphate and stir well to remove suspended moisture particles.
9. Determine transmission at 485 nm against blank prepared in similar manner.
10. Calculate concentration of iron in sample by plotting value on standard curve.
11. Report amount of iron as mg/100 g sample.

E. Estimation of Copper

Requirements : Standard copper solution – Dissolve 318 mg of pure copper in nitric acid and evaporate to dryness on steam bath. Add sufficient water and few drops of acetic acid and again evaporate to dryness. Re-dissolve in water using few drops of acetic acid and dilute to 1000 ml.

Sodium thiosulphate solution (0.01N) – Dissolve 24.82 g of sodium thiosulphate in 1000 ml carbon dioxide free water and standardize against standard copper solution.

Starch indicator – Make 1g soluble starch into paste and make up volume to 100 ml with hot water. Cool and decant clear solution and add 2 ml of chloroform to preserve.

Dilute acetic acid (1:1), Dilute nitric acid (2:5), Dilute hydrochloric acid (1:4), 1% ammonium sulphate solution, Saturated sodium thiosulphate solution, potassium iodide, sulphuric acid.

Standardization

1. Place 20 ml of standard copper solution in 250 ml volumetric flask and add excess of ammonium hydroxide and boil gently.
2. Add 10 ml of dilute acetic acid, boil and cool.
3. Add 2g potassium iodide, dissolved in water, to make final volume of 50 ml.
4. Titrate with standard sodium thiosulphate solution to light blue colour end point.
5. Add 2 ml starch indicator and continue titration until blue colour disappears.
6. Determine copper equivalent (M) to 1 ml sodium thiosulphate.

Procedure

1. Make ash with 10g of sample and dissolve in 100 ml hydrochloric acid.
2. Add 5 ml sulphuric acid and dilute solution to 200 ml and boil for one minute.
3. Add 10 ml of hot saturated sodium thiosulphate solution and boil for 5 minutes.
4. Cool and filter precipitate through Whatman No. 42 filter paper and wash six times with hot water.
5. Place folded filter paper in crucible and ignite at 500° C.
6. Treat residue with 1 ml dilute nitric acid and dry on steam bath.
7. Add 5 ml water and again dry and add 20 ml water.
8. Add excess of ammonium hydroxide and heat until copper salts are dissolved.
9. Transfer to 250 ml volumetric flask, add 10 ml of dilute acetic acid, boil and cool the solution.
10. Add 2g potassium iodide dissolved in water to make final volume of 50 ml.

11. Titrate free iodine with 0.01N sodium thiosulphate solution until blue colour appears.
12. Add 2 ml of starch indicator and continue titration until blue colour disappears.

Calculation : Copper, percent by weight = VM / 10 W

Where, V = ml of 0.01N sodium thiosulphate used for titration.

M = mg of copper equivalent to 1 ml of sodium thiosulphate.

W = Weight of prepared sample taken for test.

G. Determination of Sodium (Flame emission spectrometric Method)

Procedure

Preparation of standard solution – Weigh 1.016g of sodium chloride, transfer to 1000 ml volumetric flask and make up the volume with distilled water.

Preparation of reference solution - Pipette 15, 20, 25, 30, 40, 50 and 60 ml this solution in different 1000 ml volumetric flasks and add 900 ml water and mix. Add to each flask 10 ml of phosphorous standard solution (10.650g in 1L) and make up the volume with water and mix well.

This solution corresponds to 6,6,10,12,16,20 &24 mg of Na per litre.

Preparation of test solution – Take 1.0g of test material in 100 ml volumetric flask and dissolve in 20 ml of hot water (50 C). Cool and make up the volume with distilled water. Pipette 25 ml of the solution in 100 ml volumetric flask and make up the volume with distilled water. Pipette 10 ml of the solution in 100 ml volumetric flask, add 70 ml water, 0.2 ml of NaOH solution and 1 ml EDTA solution. Mix by swirling after each addition. Make up the volume and mix well.

Determination – Start the flame spectrometer by supplying mixture of acetylene and air to the burner. Insert the sodium filter with maximum transmittance at 589 nm. Inject reference solution starting with lowest concentration and the test solution alternatively into the flame of spectrometer using sodium filter and note the reading.

Calculation :- calcium content (%) = { (Ts – Lr / Hr – Lr) x (Q2 – Q1) + Q1} f/m

Where - Ts = reading of test solution

Lr = Lowest reading for reference solution of conc. Q1

Hr = Highest reading of reference solution of conc. Q2

Q1 = Concentration of reference solution (mg/ L) giving reading Lr

Q2 = Concentration of reference solution (mg/ L) giving reading Hr

f = Conversion factor for sodium (0.04)

m = weight of test sample (g)

H. Determination of Potassium (Flame emission spectrometric Method)

Procedure

Preparation of standard solution – Weigh 1.906g of calcium carbonate, transfer to 1000 ml volumetric flask and make up the volume with distilled water.

Preparation of reference solution - Pipette 25, 30, 35, 40, 45 and 50 ml this solution in different 1000 ml volumetric flasks and add 900 ml water and mix. Add to each flask 10 ml of .phosphorous standard solution and make up the volume with water and mix well.

This solution corresponds to 25, 30,35,40,45 &50 mg of K per litre.

Preparation of test solution – Take 1.0g of test material in 100 ml volumetric flask and dissolve in 20 ml of hot water (50 C). Cool and make up the volume with distilled water. Pipette 25 ml of the solution in 100 ml volumetric flask and make up the volume with distilled water. Pipette 10 ml of the solution in 100 ml volumetric flask, add 70 ml water, 0.2 ml of NaOH solution and 1 ml EDTA solution. Mix by swirling after each addition. Make up the volume and mix well.

Determination – Start the flame spectrometer by supplying mixture of acetylene and air to the burner. Insert the calcium filter with maximum transmittance at 623 nm. Inject reference solution starting with lowest concentration and the test solution alternatively into the flame of spectrometer using calcium filter and note the reading.

Calculation :- calcium content (%) = { (Ts – Lr / Hr – Lr) x (Q2 – Q1) + Q1} f/m

Where - Ts = reading of test solution

Lr = Lowest reading for reference solution of conc. Q1

Hr = Highest reading of reference solution of conc. Q2

Q1 = Concentration of reference solution (mg/L) giving reading Lr

Q2 = Concentration of reference solution (mg/L) giving reading Hr

f = Conversion factor for potassium (0.04)

m = weight of test sample (g)

Detection of Vitamin content in milk and milk products

1. Estimation of fat soluble vitamins

Fat isolation techniques

A. *Heat clarification* : Separate milk sample to obtain 50g cream by centrifuging at 4000 rpm for 10 minutes. Convert cream into ghee by heat clarification.

B. *Solvent extraction* : Take 100 ml milk in 500 conical flask. Add 15 ml of ammonia and shake well. Add 100 ml ethyl alcohol, 100 ml solvent ether and 100 ml petroleum ether,and shake well after each addition. Allow to stand undisturbed for 30 minutes. Decant ethereal layer into another flask. Evaporate ether to dryness at low temperature. Transfer to oven at 102° C to get pure fat.

C. *Modified Gerber method* : Take butyrometer with both ends open. Close stem side opening with acid resistance silicon cork. Add 10 ml of 90% sulphuric acid, 10.75 ml milk and 1 ml amyl alcohol. Close neck side with lock stopper and mix the contents. Centrifuge for 5 minutes to get clear fat in the column. Remove silicon cork and transfer clear fat with the help of capillary tube / syring.

Estimation of vitamin "A"

Requirements : Ethyl alcohol (95%), Pottasium hydroxide (2% & 50%), Antimony triclhoride (saturated in chloroform), Vitamin "A" standard reference with known strength (crystalline vitamin "A" acetate), anhydrous sodium sulphate.

Procedure

1. Weigh 5g of pure fat sample in a saponification flask.
2. Add 40ml ethyl alcohol and 7ml of 50% KOH solution.
3. Reflux contents for 30–40 minutes and then cool.
4. Add 150 ml distilled water and transfer contents into separating funnel.
5. Extract three times with 50ml portions of ethyl ether and collect extracts in another separating funnel.
6. Wash ether extract with 2% aqueous KOH solution and then with water till alkali free.
7. Filter ether layer through anhydrous sodium sulphate using Whatman No 40 filter paper.
8. Rinse sodium sulphate with solvent ether to extract vitamin "A"
9. Evaporate extract to dryness and dissolve residue in 5 ml chloroform.

Preparation of standard curve

1. Take known quantity of vitamin "A" acetate and carryout saponification, extraction and evaporation in similar manner as for sample.
2. Dissolve residue in known quantity (10-20ml) 0f chloroform.
3. Prepare a series of dilutions from this solution giving range from 10-50 IU / ml.
4. Take 1 ml of each dilution and add 4 ml of antimony trichloride solution.
5. Develop colour and measure absorption.
6. Plot a graph between absorption and concentration of vitamin "A".

Assay

1. Set the instrument to 100% transmission at 620 nm.
2. Take 1 ml of pure chloroform for blank and add 4 ml antimony trichloride solution in photometric cell tube.
3. In another cuvette take 1 ml sample and add 4 ml antimony trichloride solution.
4. Measure absorption within 3 seconds.
5. Plot reading in standard curve and obtain value for calculation.
6. Report result as IU of vitamin "A" per milligram sample.

   Calculation : Reading from calibrated curve = B IU (1 ml sample)

5 ml of chloroform solution of sample = B x 5 IU

1 mg sample contains (IU) = B x 5 x 100 / W

W = weight of sample (5g)

Estimation of Vitamin "E"

Requirements : Ethyl alcohol (95%), Potassium hydroxide (2% & 50%), ferric chloride (0.2% in alcohol), Vitamin "E" standard reference (pure alfa-tocopherol), benzene, hydrochloric acid, floridine XS column (12 x 30 mm) with purified absorbent, a,a` dipyridyl solution (0.5%).

Procedure

1. Weigh 5g of pure fat sample in a saponification flask.
2. Add 40ml ethyl alcohol and 7ml of 50% KOH solution.
3. Reflux contents for 30–40 minutes and then cool.
4. Add 150 ml distilled water and transfer contents into separating funnel.
5. Extract three times with 50ml portions of ethyl ether and collect extracts in another separating funnel.
6. Wash ether extract with 2% aqueous KOH solution and then with water till alkali free.
7. Filter ether layer through anhydrous sodium sulphate using Whatman No 40 filter paper.
8. Evaporate the ether extract to dryness over water bath.
9. Dissolve residue in 5ml of benzene and elute contents on floridine column with benzene.
10. Collect 25 ml elutes volume and evaporate under vacuum.
11. Dissolve residue in 10 ml ethyl alcohol to remove carotene (solution A).

Preparation of standard curve

1. Take known quantity of vitamin "E" (alfa-tocopherol) and carryout saponification, extraction and evaporation in similar manner as for sample.
2. Dissolve residue in known quantity (10-20 ml) 0f benzene.
3. Prepare a series of dilutions from this solution giving range from 20-100 μg.
4. Take 1 ml of each dilution and add 1 ml of ferric chloride solution and 1 ml of a,a'-dipyridyl solution.

5. Allow to stand for 15 minutes to develop colour and measure absorption at 530 nm.
6. Plot a graph between absorption and concentration of vitamin "E".

Assay

1. Set the instrument to 100% transmission at 530 nm.
2. Take 5 ml of solution A, from above procedure, and add 1 ml of ferric chloride solution and 1 ml of a,a'di-pyridyl solution and mix well.
3. Make up volume to 10 ml with alcohol.
4. Allow to stand for 15 minutes and measure optical density at 530 nm against blank.
5. Prepare blank in similar manner as sample.
6. Calculate concentration of Vitamin "E" from standard curve.
7. Report result as μg of vitamin "E" per milligram sample.

   Calculation : Reading from calibrated curve = Y

   10 ml of alcohol solution has = Y x 10 μg

   100 g sample contains (Vitamin "E") = Y x 10 μg /100

2. Estimation of Water soluble Vitamins

(i) Ascorbic acid

Requirements : Metaphosphoric acid- acetic acid – Dissolve 15g of pulverized sticks of glacial metaphosphoric acid an 40 ml of acetic acid and 200 ml water. Dilute to 500ml with acetic acid and filter the solution.

Standard ascorbic acid solution – Take 0.1g of reference standard ascorbic acid and dilute to 100 ml with metaphosphoric acid–acetic acid reagent.

Standard indophenol solution – Dissolve 0.05g of sodium 2,6-dichloro-benzenone indophenol in 50 ml water to which 42g of sodium bicarbonate has been added. Shake vigorously and dilute to 200 ml with water. Filter and store in amber coloured bottle.

Procedure

1. Weigh 5g of sample in 100ml volumetric flask.
2. Add 40 ml of hot water and shake well.
3. Make up volume with metaphosphoric acid–acetic acid reagent and filter.
4. Take 10 ml of filtrate and titrate with indophenol solution.
5. Carry out blank determination with 11 ml reagent along with water.
6. Calculate Vitamin C content as mg / 100g sample.

Calculation : Vitamin C percent by weight = A x B x 100 / W

Where,

A = Volume (ml) of indophenol solution used for titration.

B = Weight (mg) of ascorbic acid equivalent to 1 ml of indophenol solution.

W = Weight of sample (g)

(ii) Nicotinic acid

Requirements

Stock solution – Dissolve 50 mg of nicotinic acid reference standard in 500 ml, alcohol.

Standard solution – Dilute 2 ml of stock solution in 50 ml water to give 4 µg /ml nicotinic acid.

Sulfanilic acid solution – Take 20g of sulfanilic acid and add 170 ml water. To 1 ml of this mixture, add ammonium hydroxide solution until acid dissolves and dilute to 200 ml.

Cyanogen bromide solution – Weigh 16g of bromine in cold 200 ml volumetric flask kept in ice bath. Add 100 ml of 10% aqueous sodium cyanide slowly using a burette and make up the volume with water.

Ammonium hydroxide solution – Dilute 5 ml of ammonium hydroxide solution to 250 ml with water.

Procedure

a) Preparation of sample solution – Weigh 25g sample in 500 ml conical flask. Add 200 ml of 1N sulphuric acid, mix well and heat for 30 minutes at 100° C. Cool and adjust pH to 4.5. Dilute to 250 ml with water.

b) Preparation of sample test solution – Weigh 17g of ammonium sulphate in 50 ml graduated flask. Add 40 ml of sample solution and make up the volume with water. Filter, mix well and use 1 ml for colour development.

c) Preparation of standard test solution – Take 40 ml of nicotinic acid standard solution in a flask containing 17g ammonium sulphate. Mix well and make up volume with water. This solution contains 3.2 μg nicotinic acid per ml.

Preparation of Assay

1. Take 1 ml of standard test solution in a test tube and mark it as standard blank.
2. Take 1 ml of sample test solution in another tube and mark it as sample blank.
3. To each tube add 5 ml water, 0.5 ml ammonium hydroxide solution, 2 ml of sulfanilic acid solution and 1 drop of dilute hydrochloric acid (1:5 v/v).
4. Set colorimeter to zero absorbance at 450 nm, within 30 seconds, with each solution.
5. Similarly take 1 ml each of standard test solution and sample test solution in two test tubes.
6. Add to each tube add 0.5 ml ammonium hydroxide solution, 5 ml of cyanogens bromide solution, 2 ml of sulfanilic acid solution and 1 drop water.
7. Allow to stand for two minutes for the colour to develop.
8. Determine absorbance with sample blank set at zero.
9. The nicotinic content is proportional to absorbance if standard and sample test solutions nave same absorbance.
10. Calculate Nicotinic acid content as mg/ 100g sample.

(iii) Riboflavin (Vitamin B2)

Requirements : Stock solution – Dissolve 50mg of riboflavin reference standard in 500ml of 0.02N acetic acid over steam bath. Take 50 ml of this solution and dilute to 500 ml with 0.02N acetic acid. One ml of this solution is equivalent to 10 μg of riboflavin.

Standard solution - Dilute 10 ml of stock solution to 100 ml with water. One ml of this solution represents 1 μg of riboflavin.

Potassium permanganate – Dissolve 4g of potassium permanganate crystals in 100 ml distilled water.

Hydrogen peroxide solution(3%), Sodium hydroxide solution (4%), HCl (0.1N), Dilute HCl (1:1) and acetic acid.

Procedure

1. Weigh 5g of ash and dissolve in 50 ml of 0.1N HCl.
2. Heat mixture in autoclave for 20 minutes and cool.
3. Adjust pH to 6.0-6.5 using sodium hydroxide solution.
4. Dilute mixture to known volume with diluted HCl so that the final mixture contains more than 0.1 μg riboflavin.
5. Filter and collect clear filtrate.
6. Take an aliquot of filtrate and dilute to known volume so that the final mixture contains 0.1 μg riboflavin.
7. Take 10 ml of sample solution to each of two tubes.
8. To one tube add 1 ml of standard riboflavin solution and mix well.
9. To another tube add 1 ml of water and mix well.
10. To each tube add 1 ml of glacial acetic acid and 0.5 ml of potassium permanganate solution and allow to stand for 2 minutes.
11. To each tube add 0.5 ml of hydrogen peroxide solution and shake vigorously.
12. Measure florescence of sample solution at 565 nm. Note as reading "A".

13. Measure florescence of solution containing water at 565 nm. Note as reading "B".

14. Add to second tube 20 mg of powdered sodium hydrosulphite and measure floroscence. Note as reading"C".

15. Express result as Riboflavin mg / 100 g sample.

    Calculation : Riboflavin (mg / ml) of final solution = (B – C / A – B) 1/10 x 1/100

(iv) Thiamine content

Requirements : Stock solution – Take 50-60mg of thiamine hydrochloride reference standard and dissolve in small quantity of 20% ethyl alcohol, adjust pH to 4.3 with HCl and make up volume to 500 ml with acidified alcohol. Dilute 100 ml of thiamine hydrochloride stock solution to 1000 ml with 20% ethyl alcohol. One ml of this solution is equivalent to 10 µg of thiamine hydrochloride.

Standard solution – To 10 ml of above solution add 50 ml of 0.1N HCl and digest at 95° C for 30 minutes. Cool and dilute to 100 ml with 0.1N HCl to give 1 µg / ml.

Sodium acetate solution (2N) – Dissolve 275g sodium acetate in 1000 ml water.

Bromocresol green indicator – Dissolve 0.1g in 2.8 ml of 0.05N sodium hydroxide solution by titrating in agate mortar and dilute to 200 ml with water.

Acidified potassium chloride solution – To 1000 ml of neutral potassium chloride solution add 8.5 ml of concentrated HCl.

Oxidizing agent – Mix 4 ml of 1% potassium ferricyanide solution with 15% of sodium hydroxide solution and dilute to 100 ml.

Enzyme solution (10%) amylase P /Polidase S / Clarase potent to diastase and phosphorolytic activity, prepared fresh.

Purified base exchange silicate – Wash 100-500g base exchange silicate first with 3% hot acetic acid for 10-15 minutes followed by washing with potassium chloride solution and finally with hot distilled water.

Procedure

1. Weigh 5g sample and dissolve in 50 ml of 0.1N HCl.
2. Digest contents for 30 minutes over steam bath.
3. Cool and dilute with 0.1N HCl to known volume so that each ml of solution contains 0.2-0.5 µg of thiamine.
4. Take an aliquot containing 10-25 µg thiamine and dilute to 65 ml with 0.1N HCl.
5. Adjust pH to 4.0-4.5 with sodium acetate solution using bromocresol as indicator.
6. Add 5 ml of 10% enzyme solution and incubate at 45-50° C for 3 hours.
7. Cool and dilute to 100 ml with 0.1N HCl and filter.
8. Pass through base exchange silicate tube a aliquot of filtrate containing 5 µg of thiamine.
9. Wash column with 3 portions of boiling water.
10. Elute thiamine by passing 4.0 - 4.5 ml of boiling acidified potassium chloride solution.
11. Collect elute in 25 ml graduated flask, cool and dilute upto the mark with acidified potassium chloride solution. (Assay solution)
12. Take four 40 ml tubes.
13. To one of them add 5 ml of standard thiamine hydrochloride solution.
14. Immediately add 3 ml of oxidizing solution and 13 ml of iso-butyl alcohol and mix well. (oxidized standard solution)
15. Treat another tube (standard blank) in similar manner using 4% sodium hydroxide solution in place of oxidizing solution.
16. To third and fourth tube add 5 ml of assay solution and treat in similar manner.
17. The third tube containing standard solution shall be oxidized (oxidized standard solution).

18. In the fourth tube add 4% sodium hydroxide solution in place of oxidizing solution. (assay blank)
19. Centrifuge tubes at low speed and decant 10 ml of upper layer from each tube.
20. Measure fluorescence for each tube at 440-565 nm.
21. Express result as mg of thiamine hydrochloride per 100g sample.

Observations

Measure fluorescence from oxidized assay solution and mark as reading "A"

Measure fluorescence from assay solution, treated with NaOH, and mark as reading "b"

Measure fluorescence from oxidized standard solution and mark as reading "S"

Measure fluorescence from oxidized standard solution, treated with NaOH, and mark as reading "d"

Calculation : Thiamine hydrochloride (μg) in 5 ml solution = 1.0 (A – b) / (S – d)

❑❑❑

Chapter–5

Quality Assessment of Milk Products

India has emerged as the largest producer of milk in the World. This has boosted the confidence of producers, planners, dairy entrepreneurs and scientist alike. India has poised to play a major role in global dairy market in coming decade. The rapid growth of science and technology has positive impact on our life style and with strong trend of urbanization several demographic changes are likely to occur. There going to be ever increasing demand on the role of milk and milk products for human health. Accordingly, the demand for higher standards of milk products in terms of hygiene and safety of dairy products is to be followed. The competitiveness for the export of milk products among different dairy industry has increased in recent times due to GATT agreement. The biggest challenge before Indian dairy industry is the quality and safety of milk products indented for human consumption. Legal standards are made to have control over quality of food offered for sale and to safe guard the consumer form health hazards. It is therefore, imperative to determine different parameters that define product quality and safety. The purpose is to ensure production and distribution of products within well defined and accepted standards. This chapter deals with various parameters of product testing in order to maintain requisite prescribed standards.

Aim :- Analysis of Cheese, Paneer, Channa and Khoa

Tests are performed on these products to have control over meeting the legal standards prescribed by Prevention of Food Adulteration Act with respect to minimum fat content and fat on dry matter basis.

Sampling : Cheese and Paneer – Remove a small wedge or several plugs from the outside to the center of the block. Trim the outer surface lightly and cut the remaining portion into cubes. Chop, mince or grind the cubes finely into a uniform mass. Pack it into a screw cap bottle and keep in freezer until use.

1. Fat test

(a) Gerber method

Requirements: - Cheese butyrometer (0-40%), lock stopper, butyrometer stand, stopper key, centrifuge machine, Gerber acid, amyl alcohol, water bath

Procedure

1. Weigh 3 g of grated sample in butyrometer cup and place it into the butyrometer.
2. Add 5-6 ml of luke warm water followed by 10 ml of Gerber acid, in 2-3 portions into the butyrometer using automatic measure.
3. Add 1.0 ml of amyl alcohol using automatic measure. Clean & dry the neck of butyrometer with cotton if it is wet.
4. Stopper the butyrometer with a lock stopper, using the key, and shake the contents well.
5. Keep the butyrometer in water bath at 65° C for 5 min.
6. Place the butyrometer in Gerber centrifuge, balance the machine and centrifuge the contents for 5 min. At 1100 – 1200 rpm.
7. Keep the butyrometer in water bath at 65° C for 5 min.
8. Read the fat content (colour less column) directly on the butyrometer stem.

(b) Gravimetric method

Requirements :- Rose-Gottleib tube / Majonnier flask, fat evaporating dish, rubber stoppers, ammonia solution, diethyl ether, petroleum ether, ethyl alcohol, centrifuge machine, hot plate, oven, desiccator, weighing balance (electronic-top loading)

Procedure

1. Weigh 1 - 2 g of well-mixed sample into the extraction tube.
2. Add 2-3 ml of water and 5 ml concentrated HCl and mix well by gentle swirling motion.
3. Add 10 ml of alcohol and again mix well.
4. Add 25 ml of diethyl ether, stopper the tube with cork and shake vigorously for one minute.
5. Remove the cork and add 25 ml of petroleum ether, stopper and shake well for 30 seconds.
6. Allow the tube to stand until ethereal layer is separated clearly from the aqueous layer (for Rose-Gottlieb method).
7. In case of Majonnier method- centrifuge the Majonnier flask for 10 minutes.
8. Decant the clear ethereal layer into a previously weighed aluminium fat dish.
9. Repeat extraction with 15 ml portions each of diethyl ether and petroleum ether, twice to completely extract fat from the sample.
10. Transfer the ethereal layers from each extraction into the fat dish.
11. Place the fat dish on a hot plate (60° – 65° C) and allow the ether layer to evaporate slowly.
12. Dry the residual fat at 98° C in an oven and then allow the fat dish to cool by placing in a desiccators.
13. Weigh the fat dish and note the weight.

14. Repeat heating, cooling and weighing until successive weighing do not differ by more than 1 mg.
15. Calculate amount of fat by difference of weight in fat dish before and after extraction (reading A).
16. Divide reading A by the weight of sample and multiply by 100 to get percent fat in sample.

2. Moisture and Total solid content

Requirements: - Aluminium moisture dish, boiling water bath / hot plate, weighing balance, hot air oven, desiccator, tong.

Procedure

1. Weigh accurately a clean and dry empty dish containing sand and glass rod. (A)
2. Transfer 3-5 g of sample into the dish and note the weight. (B)
3. Place the dish on hot plate and allow the water to evaporate.
4. Place the dish in hot air oven at 100° C for 3 hours.
5. Transfer the dish to a desiccator and allow cooling for 30 minutes.
6. Weigh the contents and note the weight. (C)
7. Repeat heating and cooling until difference in two successive weights do not exceed 0.5 mg.

Observations

Weight of empty dish = a g

Weight of empty dish + sample = b g

Weight of dish after drying = c g

Calculations

Weight of sample (A) = (Reading b – Reading a)

Weight of solids (B) = (Reading c – Reading a)

Percent moisture (Z) = (b – c / A) x 100

Percent Total solids = 100 - Z

3. Total Protein content

Requirements: - Kjeldahl flask, round bottom flask, conical flask, burette, digestion heater, distillation unit, conc. sulphuric acid, copper sulphate, potassium sulphate, N/10 NaOH, 50% NaOH, N/10 sulphuric acid, mixed indicator (methyl red and 0.1% methylene blue in ethanol).

Procedure

1. Weigh accurately 0.2 g well mixed sample in a clean and dry Kjeldahl flask.
2. Add 25 ml of concentrated sulphuric acid, 0.2 g of copper sulphate and 10 g of potassium sulphate.
3. Place the flask on digestion heater under fume chamber and heat gently to boil until contents are clear, and then for another 2 hours.
4. Allow the liquid to cool and dilute with 200 ml of distilled water.
5. Fit the flask with micro-kjeldahl distillation unit.
6. Carefully add 75-80 ml of 50% NaOH down the neck of the flask to form a layer under the acid.
7. Take 50 ml of N/10 sulphuric acid in a 250 ml conical flask and place it under the condenser so that the tip of the condenser is dipped in the liquid.
8. Start heating and distil the contents until all ammonia has passed over as indicated by collection of 150 ml distillate.
9. Stop heating and remove the conical flask. Wash the condenser tip to remove all traces of condensate.
10. Titrate the contents against N/10 NaOH using mixed indicator.
11. Carry out blank determination by using 0.5 g pure sucrose in place of milk.

Observations: - Record the titer volume for original and blank experiments.

Sl.No.	Initial reading	Final reading	Vol. Of alkali used
1.			
2.			
3.			

Calculation: - Percent total nitrogen by weight = 1.4 (B-A) N / W

Percent Total protein = Percent total nitrogen x 6.38

4. Soluble protein

Quirements: Kjeldahl flask, round bottom flask, conical flask, burette, digestion heater, distillation unit, water bath, volumetric flask, pastle-morter

Sharpes extraction solution

Stock solution – Take 136.1g sodium acetate, 47.0g sodium chloride, 8.9g calcium chloride and 57.5 ml glacial acetic acid in 1000 ml volumetric flask and makeup the volume with distilled water.

Working solution – Take 250 ml of stock solution in 1000 ml volumetric flask and makeup the volume with distilled water.

Procedure

1. Accurately weigh 3.0g cheese and take in a pastle-morter.
2. Add small amount of Sharpes extraction solution and grind to a paste.
3. Transfer the contents to a 100 ml volumetric flask and makeup the volume with Sharpes solution.
4. Keep the flask in water bath at 50° – 55° C for 1 hour and occasionally shake the contents.
5. Filter through No. 42 filter paper and collect the filtrate.
6. Take 10 ml of filtrate in a Kjeldhal flask and digest the contents as incase of total protein content.

Calculation : Percent soluble nitrogen = 10 x 14.077 x (V-B) x N x 100 / W

Percent soluble protein = Percent soluble nitrogen x 6.38

5. Total ash content

Requirements: - Single pan balance, muffle furnace, desiccator, silica crucible, tong, hot plate.

Procedure

1. Weigh the empty and dried silica crucible and note the weight.
2. Accurately weigh 1.0 g if sample and note the weight.
3. Evaporate the sample to dryness on a hot plate.
4. Place the crucible in a pre-heated muffle furnace and heat the contents at 550° C until ash is free from carbon.
5. Cool the crucible by placing in desiccator.
6. Weight the crucible containing the ash.
7. Continue heating, cooling and weighing until the difference in weight is not more than 0.1g.

Observations

1. weight of empty crucible = W g
2. weight of crucible with sample = W1 g
3. weight of crucible after drying = W2 g
4. weight of sample = W1 – W g (A)
5. weight of ash = W2 – W g (B)

Calculation: - Percent ash by weight = B / A x 100

6. Acidity

Requirements: - Single pan balance, conical flask (100 ml), burette, beaker (100ml), pestle-mortar etc.

Procedure

1. Weigh about 2g of well mixed sample in a pestle-mortar.
2. Add 20 ml of distilled water and make a fine paste.
3. Transfer the contents in a beaker or conical flask.
4. Add 2 drops of 0.5% phenolphthalein solution.
5. Titrate against N/10 sodium hydroxide until faint pink colour persists.

   Observations : Amount of sample taken = Wg

   Volume of NaOH used = V ml

   Calculations : Percent Lactic acid =(V x 0.009 / W) 100

7. Salt content

Requirements : Weighing balance, pestle-mortar, conical flask (100 ml), burette, beaker, potentiometer with electrode for chloride estimation, standard silver nitrate solution, 4N nitric acid etc.

Procedure I

1. Weigh about 2g of well mixed sample of Cheese in a pestle-mortar.
2. Add small amount of distilled water and make a paste.
3. Add further 20 ml of distilled water and stir well.
4. Transfer the contents in a beaker or conical flask.
5. Add 1 ml potassium chromate indicator.
6. Titrate against N/20 silver nitrate solution until chocolate red colour persists.

   Calculation : Percent salt (NaCl) = (V x 0.00292 / W) 100

   Where V = Volume of silver nitrate used.

   W = Weight of cheese taken

Procedure II (Direct Reading)

1. Weigh about 10g of well mixed sample in a pastle-motter.
2. Add small amount of hot distilled water (40° C) and make a paste.
3. Transfer the contents in 250 ml volumetric flask.
4. Add further 20 ml portions of distilled water, stir well, transfer contents to volumetric flask and makeup the volume.
5. Mix the contents well and filter through Whatman filter paper No. 1.
6. Collect filtrate in a clean beaker.
7. Transfer 25 ml of filtrate in 100 ml conical flask and add 2 ml of 2% potassium chromate indicator
8. Titrate against 0.17 N silver nitrate solution until first perceivable colour change.

   Calculation : Percent salt (NaCl) = Volume of silver nitrate used for titration.

Procedure III

1. Weigh 2-5g of test sample into a beaker.
2. Add 30 ml of hot distilled water to suspend the material.
3. Add 2-3 ml of 4N nitric acid solution.
4. Place the beaker onto magnetic stirrer and start.
5. Place measuring and reference electrode into the solution.
6. Titrate against 0.12N silver nitrate solution until the end point has reached.
7. Run a blank determination using reagents only.

   Calculation:- % NaCl = {(V2 – V1) x N x 5.84} / W

   Where V1 = Volume of silver nitrate used in blank

   V2 = Volume of silver nitrate used for sample

   N = Normality of silver nitrate

   W = weight of sample

8. Total volatile Fatty Acids

Requirements : Weighing balance, pestle-mortar, kjeldahl flask (50 ml), burette, beaker, etc.

Procedure

1. Weigh about 10g of well mixed sample of Cheese and transfer in a pestle-mortar.
2. Add 25 ml of distilled of 10% sulphuric acid and grind the contents well.
3. Transfer the contents in 800 ml kjeldahl flask .
4. Add further 25 ml of 10% sulphuric acid to wash mortar and transfer contents to kjeldahl flask.
5. Add 35g of magnesium sulphate, few glass beads and shake the flask well.
6. Add 250 ml of glass distilled water, mix well and subject to steam distillation.
7. Collect 280 ml of distillate and titrate against N/10 NaOH using phenolphthalein indicator to the pink colour end point. (A)
8. Wash the condenser with 25 ml of neutral alcohol and titrate contents against N/10 NaOH using phenolphthalein indicator to the pink colour end point. (B)
9. Add reading A and B

Result : Volatile Fatty Acids = Number of ml of N/10 NaOH used per 10g Cheese

Aim :- Lactometric determination of Solids content in sweetened flavoured milk / lassi

Lactometers are used for rapid determination of specific gravity of liquids. Estimation of total solids in flavoured milk and lassi by lactometer is not possible as addition of sugar enhances specific gravity. For lactometric determination a modified test has been devised.

Requirements :- Lactometer (ISI), lactometer jar, thermometer

Procedure

1. Dilute the sample of flavoured milk (1:5-1:6) and lassi (1:2) to bring down their sp. Gr. to that of milk.
2. Temper the diluted sample to a temperature close to temperature of graduation of lactometer.
3. Pour sufficient sample into the lactometer jar to allow free floating of lactometer.
4. Place the lactometer in the milk and allow it to float till it stops and assumes a constant level.
5. Record the lactometer reading and temperature of milk at the same time.
6. Take another reading by flapping the top of the lactometer stem and when it again assumes constant level.
7. Correct lactometer reading at 27°C by applying temperature correction.
8. Calculate solids-not-fat (SNF) content using the given formulas.

Calculations (i) For Sweetened flavoured milk

a. *Using Quevennue lactometer* - % TS = (0.223 CLR + 1.18F +0.73) x 2

b. *Using Zeal lactometer* - % TS = (0.225 CLR + 1.17F +1.00) x 2

c. *Using BIS lactometer* - % TS = 0.5 CLR + 1.2F + 0.50

% milk s.n.f. = Fat + CLR/ 10 + 5.5

Where, CLR is corrected lactometer reading and F is % fat in diluted milk sample.

(ii) For sweetened lassi

Using Zeal lactometer - % TS = (0.205 CLR + 1.05F +1.42) × 3

Where F is % fat in diluted milk sample

% milk solids. = 0.5 CLR + 1.0F + 6.0

Using BIS lactometer - % TS = 0.6 CLR + 1.2F + 4.40

Where, CLR is corrected lactometer reading and F is % fat in undiluted milk sample.

Acidity in flavoured milk

The acidity of flavoured milk containing colour is difficult to determine using phenolpthaleine as indicator because of no distinct or sharp observation in end point colour change. Therefore, titration has to be carried out potentiometrically using pH meter.

Requirements :- Beaker, pH meter, standards buffer tablets, volumetric flask, magnetic stirrer, N/10 sodium hydroxide solution

Procedure

1. Calibrate the pH meter using standard or known buffered solution.
2. Take 20 ml of sample in a 50 ml beaker containing magnetic starring bar.
3. Place the beaker on a magnetic stirrer and start the stirrer.
4. Dip the electrode in the sample and titrate the contents using N/10 sodium hydroxide solution until the sample attain a pH of 8.35-8.40

Calculation:- % acidity = V/W

Where V= Volume of N/10 NaOH solution used.

W = Weight of sample

Aim :- Analysis of Dahi, Yoghurt and Srikhand

Sampling : - The representative sample for these fermented milk products are drawn by thorough mixing of product, in the container, using a spoon or spatula. The contents are moved up and down rotary motion so that the top layers and contents of lower corners of the container are well mixed and attain homogeneity.

1. Moisture and Total solid content

Requirements: - Aluminium moisture dish, boiling water bath / hot plate, weighing balance, hot air oven, desiccator, tong.

Procedure

1. Weigh accurately a clean and dry empty dish containing sand and glass rod. (A)
2. Transfer 3-5 g of sample into the dish and note the weight. (B)
3. Place the dish on hot plate and allow the water to evaporate.
4. Place the dish in hot air oven at 100° C for 3 hours.
5. Transfer the dish to a desiccator and allow cooling for 30 minutes.
6. Weigh the contents and note the weight. (C)
7. Repeat heating and cooling until difference in two successive weights do not exceed 0.5 mg.

Observations

Weight of empty dish	=	a g
Weight of empty dish + sample	=	b g
Weight of dish after drying	=	c g

Calculations

Weight of sample (A)	=	(Reading b – Reading a)
Weight of solids (B)	=	(Reading c – Reading a)
Percent Moisture (Z)	=	(b – c / A) x 100
Percent Total solids	=	100 - Z

2. Fat content

Requirements: - Rose-Gottleib tube / Majonnier flask, fat evaporating dish, rubber stoppers, ammonia solution, diethyl ether, petroleum ether, ethyl alcohol, centrifuge machine, hot plate, oven, desiccator, weighing balance (electronic-top loading)

Procedure

1. Weigh 1 - 2 g of well-mixed sample into the extraction tube.
2. Add 2-3 ml of water and 5 ml concentrated HCl and mix well by gentle swirling motion.
3. Add 10 ml of alcohol and again mix well.
4. Add 25 ml of diethyl ether, stopper the tube with cork and shake vigorously for one minute.
5. Remove the cork and add 25 ml of petroleum ether, stopper and shake well for 30 seconds.
6. Allow the tube to stand until ethereal layer is separated clearly from the aqueous layer (for Rose-Gottlieb method).
7. In case of Majonnier method- centrifuge the Majonnier flask for 10 minutes.
8. Decant the clear ethereal layer into a previously weighed aluminium fat dish.
9. Repeat extraction with 15 ml portions each of diethyl ether and petroleum ether, twice to completely extract fat from the sample.
10. Transfer the ethereal layers from each extraction into the fat dish.
11. Place the fat dish on a hot plate (60° – 65° C) and allow the ether layer to evaporate slowly.
12. Dry the residual fat at 98° C in an oven and then allow the fat dish to cool by placing in a desiccators.
13. Weigh the fat dish and note the weight.
14. Repeat heating, cooling and weighing until successive weighing do not differ by more than 1 mg.
15. Calculate amount of fat by difference of weight in fat dish before and after extraction (reading A).

16. Divide reading A by the weight of sample and multiply by 100 to get percent fat in sample.

3. Total Protein content

Requirements: - Kjeldahl flask, round bottom flask, conical flask, burette, digestion heater, distillation unit, conc. sulphuric acid, copper sulphate, potassium sulphate, N/10 NaOH, 50% NaOH, N/10 sulphuric acid, mixed indicator (methyl red and 0.1% methylene blue in ethanol).

Procedure

1. Weigh accurately 0.2 g well mixed sample in a clean and dry Kjeldahl flask.
2. Add 25 ml of concentrated sulphuric acid, 0.2 g of copper sulphate and 10 g of potassium sulphate.
3. Place the flask on digestion heater under fume chamber and heat gently to boil until contents are clear, and then for another 2 hours.
4. Allow the liquid to cool and dilute with 200 ml of distilled water.
5. Fit the flask with micro-kjeldahl distillation unit.
6. Carefully add 75-80 ml of 50% NaOH down the neck of the flask to form a layer under the acid.
7. Take 50 ml of N/10 sulphuric acid in a 250 ml conical flask and place it under the condenser so that the tip of the condenser is dipped in the liquid.
8. Start heating and distil the contents until all ammonia has passed over as indicated by collection of 150 ml distillate.
9. Stop heating and remove the conical flask. Wash the condenser tip to remove all traces of condensate.
10. Titrate the contents against N/10 NaOH using mixed indicator.
11. Carry out blank determination by using 0.5 g pure sucrose in place of milk.

Observations: - Record the titer volume for original and blank experiments.

Sl.No.	Initial reading	Final reading	Vol. Of alkali used
1.			
2.			
3.			

Calculation: - Percent total nitrogen by weight = 1.4 (B-A) N / W

Percent Total protein = Percent total nitrogen x 6.38

4. Acidity

Requirements: - Single pan balance, conical flask (100 ml), burette, beaker (100ml), pestle-mortar etc.

Procedure

1. Weigh about 2g of well mixed sample in a pestle-mortar.
2. Add 20 ml of distilled water and make a fine paste.
3. Transfer the contents in a beaker or conical flask.
4. Add 2 drops of 0.5% phenolphthalein solution.
5. Titrate against N/10 sodium hydroxide until faint pink colour persists.

Observations : Amount of sample taken = Wg

Volume of NaOH used = V ml

Calculations : Percent Lactic acid =(V x 0.009 / W) 100

Aim :- Analysis of Ice – Cream

Preparation of sample: Draw a representative sample from the batch. Transfer in a suitable container and add 1-2 drops of antifoam agent and mix well. Temper the sample by keeping in water bath at 40° – 45° C. Mix the contents well with the help of a spatula / spoon.

1. Determination of fat content

(a) Gerber method

Requirements: - Milk butyrometer (0-10%), lock stopper, butyrometer stand, stopper key, centrifuge machine, Dilute Gerber acid (70:13 for Vanilla ice cream, 94:6 for chocolate ice cream), amyl alcohol, water bath.

Procedure

1. Transfer 20-25g sample in a 100ml volumetric flask and makeup the volume with luke warm distilled water.
2. With the help of pipette transfer 10ml sample into the butyrometer.
3. Add 9.0 ml of Gerber acid (80%) into the butyrometer.
4. Add 1.0 ml of amyl alcohol using automatic measure. Clean & dry the neck of butyrometer with cotton if it is wet.
5. Stopper the butyrometer with a lock stopper, using the key, and shake the contents well.
6. Keep the butyrometer in water bath at 65° C for 5 min.
7. Place the butyrometer in Gerber centrifuge, balance the machine and centrifuge the contents for 6-7 min. At 1100 – 1200 rpm.
8. Keep the butyrometer in water bath at 65° C for 5 min.
9. Read the fat content (colour less column) directly on the butyrometer stem.

   Calculation : Percent fat = (Butyrometer reading / Wt. Of sample) x 100

(b) Gravimetric method

Requirements: - Rose-Gottleib tube / Majonnier flask, fat evaporating dish, rubber stoppers, ammonia solution, diethyl ether, petroleum ether, ethyl alcohol, centrifuge machine, hot plate, oven, desiccator, weighing balance (electronic-top loading)

Procedure

1. Weigh 2-4 g of well-mixed sample in a beaker.
2. Add 3 ml of concentrated HCl and digest on the hot plate for 30 minutes.
3. Cool and transfer contents into extraction flask.
4. Add 10 ml of alcohol and again mix well.
5. Add 25 ml of diethyl ether, stopper the tube with cork and shake vigorously for one minute.
6. Remove the cork and add 25 ml of petroleum ether, stopper and shake well for 30 seconds.
7. Allow the tube to stand until ethereal layer is separated clearly from the aqueous layer (for Rose-Gottlieb method).
8. In case of Majonnier method- centrifuge the Majonnier flask for 10 minutes.
9. Decant the clear ethereal layer into a previously weighed aluminium fat dish.
10. Repeat extraction with 15 ml portions each of diethyl ether and petroleum ether, twice to completely extract fat from the sample.
11. Transfer the ethereal layers from each extraction into the fat dish.
12. Place the fat dish on a hot plate (60° – 65° C) and allow the ether layer to evaporate slowly.
13. Dry the residual fat at 125° C in an oven for 1 hour and then allow the fat dish to cool by placing in a dessicator.
14. Weigh the fat dish and note the weight.
15. Repeat heating, cooling and weighing until successive weighing do not differ by more than 1 mg.
16. Calculate amount of fat by difference of weight in fat dish before and after extraction (reading A).
17. Divide reading A by the weight of sample and multiply by 100 to get percent fat in sample.

Calculation: Percent fat = (Wt. Of residue / Wt. Of sample) × 100

2. Total solid content

Sample preparation – Plain ice cream – Allow to soften at room temperature and mix thoroughly by stirring with spoon, egg beater or pouring.

Fruit / nut ice cream – Take 100-200g sample and fill mixer cup about 1/3rd. Allow to soften at room temperature. Start mixer and mix contents until insoluble particles are finely divided (3-5 minutes for fruit ice cream and 7-8 minutes for nut ice cream). Transfer to suitable container for weighing and store at – 15° C.

Requirements: - Aluminium moisture dish, boiling water bath / hot plate, weighing balance, hot air oven, desiccator, tong.

Procedure

1. Weigh accurately a clean and dry empty dish containing sand and glass rod. (A)
2. Transfer 3-5 g of sample into the dish and note the weight. (B)
3. Place the dish on hot plate and allow the water to evaporate.
4. Place the dish in hot air oven at 100° C for 3 hours.
5. Transfer the dish to a desiccator and allow cooling for 30 minutes.
6. Weigh the contents and note the weight. (C)
7. Repeat heating and cooling until difference in two successive weights do not exceed 0.5 mg.

Calculations: - Percent total solids = (C-A / B-A) x 100

3. Total Protein content

(a) Semi-micro Kjeldahl method

Requirements: - Kjeldahl flask, round bottom flask, conical flask, burette, digestion heater, distillation unit, conc. sulphuric acid, copper sulphate, potassium sulphate, N/10 NaOH, 50% NaOH, N/

10 sulphuric acid, mixed indicator (methyl red and 0.1% methylene blue in ethanol).

Procedure

1. Weigh accurately 0.2-0.3 g well mixed sample in a clean and dry Kjeldahl flask.
2. Add 8 ml of concentrated sulphuric acid, 2.0 g of digestion mixture (potassium sulphate & copper sulphate (50:1).
3. Place the flask on digestion heater under fume chamber and heat gently to boil until contents are clear, and then for another 2 hours.
4. Allow the liquid to cool and dilute with 200 ml of distilled water.
5. Fit the flask with micro-kjeldahl distillation unit.
6. Carefully add 25 ml of 50% NaOH down the neck of the flask to form a layer under the acid.
7. Take 25 ml of saturated boric acid in a 250 ml conical flask and place it under the condenser so that the tip of the condenser is dipped in the liquid.
8. Start heating and distil the contents until all ammonia has passed over as indicated by collection of 150 ml distillate.
9. Stop heating and remove the conical flask. Wash the condenser tip to remove all traces of condensate.
10. Titrate the contents against N/50 HCl using 0.3ml mixed indicator (100mg methyle red and 30mg methylene blue).
11. Carry out blank determination by using 0.5 g pure sucrose in place of milk.

    Calculation: Percent Protein = (0.028xVol of acid used / Wt. Of sample) × 100

(b) Formal titration method

Procedure

1. Take 4-5g sample in a 50ml conical flask.

2. Add 10 ml of distilled water and mix well.
3. Add 0.4 ml of saturated potassium oxalate solution and 1.0 ml of phenolphthalein indicator.
4. After 5 minutes titrate the contents with N/9 NaOH.
5. Add 2.0 ml neutral Formaline and hold for 5 minutes.
6. Titrate the sample using N/9 NaOH to pink colour end point.

   Calculation : Percent Protein = (Vol of NaOH used x 19.55 / Sample Wt.) x 100

4. Sucrose content (Lane-Eynon method).

Requirements: - Volumetric flask, beaker, conical flask, burette, funnel, filter stand, chemical balance, hot plate, filter paper, 10% acetic acid,

Reagents: Stock invert sugar solution – Transfer 9.5g of sucrose in 1000ml volumetric flask. Add 100ml water and 5ml concentrated HCl. Allow to stand for three days and then makeup the volume.

Standard solution – Neutralize know quantity of stock solution with NaOH using litmus paper and diluting with water to a known volume, so that not more than 15ml and less than 50ml, is required to reduce all the copper in Fehling solution taken for titration. Note concentration of sugar solution as mg/100ml. Prepare fresh.

Methylene blue – Dissolve 0.2g methylene blue in 100ml water.

Fehling solution – Mix equal volume of standard Fehling A and B solution.

Zinc acetate – Dissolve21.9g of crystalline zinc acetate in 40ml distilled water, add 3ml glacial acetic acid and makeup the volume to 100ml with distilled water.

Potassium ferrocynide solution – Dissolve 10.6g crystalline potassium ferrocynide to 100ml with distilled water.

Dilute ammonia solution – Take 10ml of concentrated ammonia and dilute to 100ml with distilled water.

Dilute acetic acid solution – Take 10ml of glacial acetic acid and dilute to 100ml with distilled water.

Standardization of Fehling solution

1. Fill burette with standard sugar solution.
2. Pipette 10ml of fehling solution into 250ml conical flask.
3. Run enough sugar solution into the flask, so that not more than 1ml will be required to complete the titration later.
4. Heat the contents to boiling for 2 minutes and then add 1ml of methylene blue indicator.
5. While the contents are boiling, add sugar solution drop by drop till blue colour disappears.

Preparation of sample solution:

(A) Solution B-I - Weigh 40g of well mixed sample and transfer to 100ml beaker. Add 50ml of hot water, mix and transfer to 200ml volumetric flask. Cool and add 5ml of diluted ammonia solution and allow to stand for 15 minutes. Add dilute acetic acid solution enough to neutralize ammonia added. Mix well and add 12.5ml of zinc acetate solution followed by addition of 12.5ml potassium ferocynide solution. Makeup the volume with distilled water. Allow to settle and filter. Mark it as solution B-I.

Solution A-I - Pipette 50ml of solution B-I into 100ml volumetric flask, add 5ml of concentrated HCl and heat to 68° C for 5 minutes. Cool and neutralize with N/10 NaOH solution.

Solution A-II and B-II - Dilute solution B-I and A-I separately, so that the volume of solution required to react with 10ml Fehling solution is between 15-50ml.

Method of titration

1. Pipette 10 ml of Fehling solution into 300ml conical flask.
2. Add all of solution B-II.
3. Gently boil the contents for 2 minutes and add 1ml methylene blue indicator.
4. Add prepared solution from the burette till blue colour diappear.

Calculation: Percent sucrose = 20a/W (F2/V2 – F1/V1)

Where,

W = weight of sample in gm.

a = weight of sucrose corresponding to 10ml Fehling solution

F2 = Dilution factor for solution A-II from solution A-I

V2 = Volume of solution A-I corresponding to 10 ml Fehling solution

F1 = Dilution factor for solution B-II from solution B-I

V1 = Volume of solution B-II corresponding to 10 ml Fehling solution

Colorimetric estimation of sucrose

Standard sucrose solution : Concentration ranging from 0.1 to 1.0mg /ml

1. Take 1ml sucrose solution.
2. Add 5ml HCl and 2ml of 0.1% resorcinol
3. Add 2ml distilled water and boil for 8 minutes.
4. Determine absorbance at 480 nm in spectrophotometer.
5. Plot a garph between concentration and absorbance.

Deproteinzation of sample

1. Take 20g sample in a clean and dry beaker.
2. Add 100ml distilled water and 1.5ml of 41% lead acetate.
3. Boil the mixture for 1-2 minutes and add 2-3 drops of alumina cream.
4. Filter through Whatman no. 42 filter paper containing 1g activated charcoal and collect filtrate.

Estimation

1. Take 5ml of charcoal treated filtrate and dilute to 100ml with distilled water.

2. Take 1ml of diluted filtrate and add to it 5ml of concentrated HCl, 2ml distilled water and 2ml of 0.1% resorcinol.
3. Boil for 5 minutes and determine absorbance at 480 nm.
4. Extrapolate absorbance on standard curve and calculate concentration of sucrose by following formula –

   Calculation : Percent sucrose = 500xC/ W

Where, C= Concentration of sucrose in mg /ml filtrate.

W = Weight of sample (20g)

5. Determination of over-run in ice-cream

Over-run is defined as the volume of ice-cream obtained in excess of the volume of ice-cream mix. It is expressed as percent over-run.

Procedure

1. Weigh a cup or cartoon containing ice-cream.
2. Transfer ice-cream into a beaker and melt at 45° C.
3. Add 1-2 drops of anti foaming agent to remove entrapped foam.
4. Fill the cup or cartoon with melted ice-cream after removing air, to the same level to which ice-cream was filled, and note the weight.
5. Also weigh an empty dry cup or cartoon.

   Calculation :- % Overrun= (W3 – W2) / (W2 – W1) x 100

   Where,

   W1 = Weight of empty cup or cartoon

   W2 = Weight of empty cup or cartoon + sample

   W3 = Weight of empty cup or cartoon + ice-cream mix

Aim :- Analysis of Evaporated and Sweetened Condensed milk

Sample preparation – Place the can of condensed milk in water bath at 45 C until sample reaches the temperature of 40°C.

Open the container and mix contents vigorously by stirring with a spoon, in a fashion to evenly mix the top layer with contents of lower corner, until sample is thoroughly mixed. Repeat stirring before drawing sample for each test.

1. Test for fat (Gravimetric method)

Requirements: - Rose-Gottleib tube / Majonnier flask, fat evaporating dish, rubber stoppers, ammonia solution, diethyl ether, petroleum ether, ethyl alcohol, centrifuge machine, hot plate, oven, desiccator, weighing balance (electronic-top loading)

Procedure

1. Weigh 2-4 g of well-mixed sample in a beaker.
2. Add 3 ml of concentrated HCl and digest on the hot plate for 30 minutes.
3. Cool and transfer contents into extraction flask.
4. Add 10 ml of alcohol and again mix well.
5. Add 25 ml of diethyl ether, stopper the tube with cork and shake vigorously for one minute.
6. Remove the cork and add 25 ml of petroleum ether, stopper and shake well for 30 seconds.
7. Allow the tube to stand until ethereal layer is separated clearly from the aqueous layer (for Rose-Gottlieb method).
8. In case of Majonnier method- centrifuge the Majonnier flask for 10 minutes.
9. Decant the clear ethereal layer into a previously weighed aluminium fat dish.
10. Repeat extraction with 15 ml portions each of diethyl ether and petroleum ether, twice to completely extract fat from the sample.
11. Transfer the ethereal layers from each extraction into the fat dish.
12. Place the fat dish on a hot plate (60° – 65° C) and allow the ether layer to evaporate slowly.

13. Dry the residual fat at 125° C in an oven for 1 hour and then allow the fat dish to cool by placing in a dessicator.
14. Weigh the fat dish and note the weight.
15. Repeat heating, cooling and weighing until successive weighing do not differ by more than 1 mg.
16. Calculate amount of fat by difference of weight in fat dish before and after extraction (reading A).
17. Divide reading A by the weight of sample and multiply by 100 to get percent fat in sample.

    Calculation: Percent fat = (Wt. Of residue / Wt. Of sample) × 100

2. Total solid content

Requirements: - Aluminium moisture dish, boiling water bath / hot plate, weighing balance, hot air oven, desiccator, tong.

Procedure

1. Weigh accurately a clean and dry empty dish containing sand and glass rod. (A)
2. Transfer 3-5 g of sample into the dish and note the weight. (B)
3. Place the dish on hot plate and allow the water to evaporate.
4. Place the dish in hot air oven at 100° C for 3 hours.
5. Transfer the dish to a desiccator and allow cooling for 30 minutes.
6. Weigh the contents and note the weight. (C)
7. Repeat heating and cooling until difference in two successive weights do not exceed 0.5 mg.

   Calculations :- Percent total solids = (C-A / B-A) x 100

3. Sucrose content (Lane-Eynon method)

Requirements: - Volumetric flask, beaker, conical flask, burette, funnel, filter stand, chemical balance, hot plate, filter paper, 10% acetic acid,

Reagents: Stock invert sugar solution – Transfer 9.5g of sucrose in 1000ml volumetric flask. Add 100ml water and 5ml concentrated HCl. Allow to stand for three days and then makeup the volume.

Standard solution – Neutralize know quantity of stock solution with NaOH using litmus paper and diluting with water to a known volume, so that not more than 15ml and less than 50ml, is required to reduce all the copper in Fehling solution taken for titration. Note concentration of sugar solution as mg/100ml. Prepare fresh.

Methylene blue – Dissolve 0.2g methylene blue in 100ml water.

Fehling solution – Mix equal volume of standard Fehling A and B solution.

Zinc acetate – Dissolve 21.9g of crystalline zinc acetate in 40ml distilled water, add 3ml glacial acetic acid and makeup the volume to 100ml with distilled water.

Pottasium ferrocynide solution – Dissolve 10.6g crystalline potassium ferocynide to 100ml with distilled water.

Dilute ammonia solution – Take 10ml of concentrated ammonia and dilute to 100ml with distilled water.

Dilute acetic acid solution – Take 10ml of glacial acetic acid and dilute to 100ml with distilled water.

Standardization of Fehling solution

1. Fill burette with standard sugar solution.
2. Pipette 10ml of Fehling solution into 250ml conical flask.
3. Run enough sugar solution into the flask, so that not more than 1ml will be required to complete the titration later.
4. Heat the contents to boiling for 2 minutes and then add 1ml of methylene blue indicator.
5. While the contents are boiling, add sugar solution drop by drop till blue colour disappears.

Preparation of sample solution

Solution B-I - Weigh 40g of well mixed sample and transfer to 100ml beaker. Add 50ml of hot water, mix and transfer to 200ml volumetric flask. Cool and add 5ml of diluted ammonia solution and allow to stand for 15 minutes. Add dilute acetic acid solution enough to neutralize ammonia added. Mix well and add 12.5ml of zinc acetate solution followed by addition of 12.5ml potassium ferocynide solution. Makeup the volume with distilled water. Allow to settle and filter. Mark it as solution B-I.

Solution A-I - Pipette 50ml of solution B-I into 100ml volumetric flask, add 5ml of concentrated HCl and heat to 68° C for 5 minutes. Cool and neutralize with N/10 NaOH solution.

Solution A-II and B-II - Dilute solution B-I and A-I separately, so that the volume of solution required to react with 10ml Fehling solution is between 15-50ml.

Method of titration

1. Pipette 10ml of Fehling solution into 300ml conical flask.
2. Add all of solution B-II.
3. Gently boil the contents for 2 minutes and add 1ml methylene blue indicator.
4. Add prepared solution from the burette till blue colour disappear.

Calculation: Percent sugar = 20a/W (F2/V2 – F1/V1)

Where,

W = weight of sample in gm.

a = weight of sucrose corresponding to 10ml Fehling solution

F2 = Dilution factor for solution A-II from solution A-I

V2 = Volume of solution A-I corresponding to 10 ml Fehling solution

F1 = Dilution factor for solution B-II from solution B-I

V1 = Volume of solution B-II corresponding to 10 ml Fehling solution.

4. Acidity

Requirements: - Single pan balance, conical flask (100 ml), burette, beaker (100ml), pastle-motar etc.

Procedure

1. Weigh about 10g of well mixed sample in a suitable dish.
2. Add 30 ml of lukewarm distilled water and mix well.
3. Add 1 ml of 0.5% phenolphthalein indicator solution.
4. Titrate against N/10 sodium hydroxide until faint pink colour persists. Complete titration in 20 seconds.

Observations : Amount of sample taken = Wg

Volume of NaOH used = V ml

Calculations : Percent Lactic acid =(V x 0.009 / W) 100

5. Protein content

Requirements: - Kjeldahl flask, round bottom flask, conical flask, burette, digestion heater, distillation unit, conc. sulphuric acid, copper sulphate, potassium sulphate, N/10 NaOH, 50% NaOH, N/10 sulphuric acid, mixed indicator (methyl red and 0.1% methylene blue in ethanol).

Procedure

1. Weigh accurately 0.5-1.0 g well mixed sample in a clean and dry Kjeldahl flask.
2. Add 8 ml of concentrated sulphuric acid, 2.0 g of digestion mixture (potassium sulphate & copper sulphate (50:1).
3. Place the flask on digestion heater under fume chamber and heat gently to boil until contents are clear, and then for another 2 hours.
4. Allow the liquid to cool and dilute with 200 ml of distilled water.
5. Fit the flask with micro-kjeldahl distillation unit.

6. Carefully add 25 ml of 50% NaOH down the neck of the flask to form a layer under the acid.
7. Take 25 ml of saturated boric acid in a 250 ml conical flask and place it under the condenser so that the tip of the condenser is dipped in the liquid.
8. Start heating and distil the contents until all ammonia has passed over as indicated by collection of 150 ml distillate.
9. Stop heating and remove the conical flask. Wash the condenser tip to remove all traces of condensate.
10. Titrate the contents against N/50 HCl using 0.3ml mixed indicator (100mg methyl red and 30mg methylene blue).
11. Carry out blank determination by using 0.5 g pure sucrose in place of milk.

Calculation: Percent Protein = (0.028xVol of acid used / Wt. Of sample) x 100

Aim :- Analysis of ghee, butter-oil and Anhydrous butter-oil

Fats and oils exhibit certain physical and chemical characteristics. The nature, proportion and manner of distribution of glycerol esters determine character of fat type. Milk fat has specific fat constants which are different than other fats. The tests are helpful in detecting adulterations and to maintain legal standards.

1. Moisture test

Requirements : Moisture dish, Desiccators, hot air oven, weighing balance.

Procedure

1. Weigh accurately a clean and dry empty dish.(A)
2. Transfer 10 g of sample into the dish and note the weight. (B)
3. Place the dish on hot plate and allow the water to evaporate.
4. Place the dish in hot air oven at 105° C for 1 hours.
5. Transfer the dish to a desiccator and allow cooling for 30 minutes.

6. Weigh the contents and note the weight. (C)
7. Repeat heating and cooling until difference in two successive weights do not exceed 0.5 mg.

Observations

Weight of empty dish	=	A g
Weight of empty dish + sample	=	B g
Weight of dish after drying	=	C g

Calculations

Weight of sample (X) = Reading B – Reading A

Weight after drying (Y) = Reading C – Reading A

Percent moisture = B – C / B - A x 100

2. Free fatty acid (FFA) OR Acidity

Acidity in ghee is expressed as percentage of free fatty acid calculated as Oleic acid as it is the main component of FFA. The maximum permissible limit is 3.0%.

(a) Titration Method

Requirements : Beaker, flask, N/10 NaOH, alcohol, burette, Phenolphthalein indicator.

Procedure

1. Weigh ~10 g of sample in a conical flask.
2. Add 50 ml of neutral alcohol and boil the contents.
3. Add few drops of indicator and titrate the contents, while boiling, using N/10 NaOH to pink color end point.
4. Note volume of NaOH used.
5. Express result in % free fatty acid (FFA).

   Calculation : % FFA = Vol. of NaOH used x 28.2 / 9 / wt. of sample

(b) Colorimetric Method

Requirements : Beaker, Conical flask, Centrifuge tubes (15ml), Clinical centrifuge, colorimeter.

Reagents :- Cupric acid pyridine (CAP) reagent adjusted to pH 6.0-6.2, Benzene.

Preparation of standard curve

Stock solution – Dissolve 6.34ml of pure Oleic acid in 100 ml benzene.

Working solution:- Pipette out 1 ml of stock solution in a 100 ml volumetric flask and make up the volume with benzene.

(a) Pipette 0.5: 1.0; 1.5, 2.0, 2.5, 3.0 and 3.5 ml of working solution in marked centrifuge tubes.

(b) Add 2.0 ml of CAP reagent in all the tubes, stopper and shake contents vigorously.

(c) Centrifuge at 5000 rpm for 5 minutes.

(d) Take supernatant and observe optical density (OD) in a colorimeter using red filter.

(e) Adjust colorimeter to 0.0 OD with benzene.

(f) Draw a graph between micromoles concentration and optical density.

Procedure

1. Weigh accurately 5.0g of fat in 25 ml volumetric flask and makeup the volume with benzene.
2. Take 1.0 ml of this solution in a glass stoppered centrifuge tube.
3. Add 2.0 ml of CAP reagent and 9.0 ml of benzene.
4. Shake vigorously and centrifuge at 5000 rpm for 5 minutes.
5. Take supernatant and observe optical density as for standard curve.
6. Plot the reading in standard graph and find out the concentration of FFA in sample.

7. Multiply the reading with 250 to calculate percent FFA in the sample.

3. B.R. Value

Requirements : Beaker, glass rod, butyrorefractometer, water bath.

Procedure

1. Melt the sample at 40° C.
2. Place one drop of sample onto butyrorefractometer lens.
3. Maintain temperature at 40° C by circulating hot water through the water bath.
4. Observe the reading on instrument scale directly and make note.
5. The BR reading for milk fat should be in the range of 40-43.

4. Reichert-Meissl Value (RM value)

It is defined as the number of milliliters of 0.1 N aqueous alkali solution required to neutralize the water soluble steam volatile fatty acids distilled from 5.0g of fat under precise conditions.

Requirements : RM flask, burrett, pipette, measuring cylinder, funnel, filter paper, distil head, burner., distillation apparatus.

Reagents : Glycerol (98%), NaOH (50%), N/9 NaOH solution, ethyl alcohol, dilute sulphuric acid (40 ml to neutralize 2 ml of 50% NaOH), phenolphthalein indicator (0.5%).

Procedure

1. Weigh ~5g of sample in a Reichert-Meissl flask.
2. Add 20g glycerol and 2 ml of 50% NaOH.
3. Heat the contents with stirring till the liquid is cleared and cool the contents to room temperature.
4. Add 93 ml of boiled distilled water before the soap is solidified.

5. Add few pieces of glass beads followed by 50 ml of diluted sulphuric acid and immediately connect the flask with distillation apparatus.
6. Heat the contents, without boiling, until insoluble acids are completely melted.
7. Increase the flame and start distillation and collect 110 ml of the distillate within 20 minutes in a marked flask.
8. Remove the flask containing distillate and discontinue heating. Cool the distillate to 15° C for 10 minutes using ice cold water.
9. Mix the contents and filter through Whatman No. 4 filter paper, reject first few drops and collect 100 ml filtrate.
10. Titrate the filtrate with N/9 NaOH using phenolphthalein as indicator.
11. Carry out blank in the same manner.

Calculation : 6.11 x vol. of NaOH used / wt. of sample

5. Peroxide value (PV)

It is an index of oxidative rancidity and is defined as number of milliliters of 0.002 N sodium thiosulphate required to neutralize the iodine librated from 1.0g of fat. It is expressed in terms of mill moles.

Requirements : Conical flask, rubber stopper with hole, balance, burette, water bath and pipette.

Reagents : Acetic acid – chloroform solvent mixture (2:1), 0.002 N sodium thiosulphste, 1% starch solution (freshly prepared).

Procedure

1. Weigh accurately 1.0 + 0.01g of sample in 150 ml conical flask.
2. Add 20.ml of solvent mixture and heat the contents on boiling water bath for 30 seconds.
3. Cool to room temperature and add 30 ml distilled water.

4. Titrate against 0.002 N sodium thiosulphate solution using starch as indicator.

   Calculation : Divide the volume of sodium thiosulphate used with the weight of sample to get PV.

6. Total Cholesterol

Declaration of total cholesterol content of ghee samples have now become mandatory to be displayed on the package as nutritional facts and hence, its determination has become important under new Indian food safety law, 2006.

Requirements : Conical flask, glass test tubes, glass stopper, balance, spectrophotometer/ colorimeter, water bath and pipette.

Reagents : Liebergmann-Burchard (LB) reagent – Mix 1.0ml of concentrated sulphuric acid in chilled 20.0ml acetic anhydride and keep solution at 0° C for 27 minutes. (to be freshly prepared and used).

(a) Direct method

Preparation of standard curve

1. Prepare 0.1% solution of pure cholesterol in chloroform.
2. Pipette aliquots of 0.5, 0.6, 0.7, 0.8, 1.0, 1.1 and 1.2 ml in a graduated tube and makeup the volume to 3.0ml with chloroform.
3. To each tube add 4.0ml of LB reagent and mix well.
4. Incubate the contents at 25° C for 12 minutes in water bath.
5. Quickly take optical density at 650 nm.
6. Plot a graph between concentration and optical density.

Estimation

1. Take 5.0g sample and dissolve in 25ml chloroform.
2. Pipette 1.0ml of this solution in a 15ml test tube.
3. Add 2.0ml chloroform and 4.0ml LB reagent.

4. Incubate at 25° C for 12 minutes and measure O.D. at 650 nm.
5. Prepare a blank using 3.0ml chloroform and 4.0ml LB reagent.
6. Calculate concentration from calibration curve.
7. Report concentration of cholesterol as mg per 100g sample

Calculation

2ml solution gives concentration = A mg cholesterol

25ml solution gives concentration = A x 25 / 2 = B mg cholesterol

(this represent mg cholesterol in 5g sample)

Total cholesterol (mg/100g) = Bx100/5

(b) Indirect method

(i) Extraction of unsaponifiable matter

1. Weigh 5g of ghee sample in a saponification flask.
2. Add 40ml ethyl alcohol and 7ml of 50% KOH solution.
3. Reflux contents for 30–40 minutes and then cool.
4. Add 150 ml distilled water and transfer contents into separating funnel.
5. Extract three times with 50ml portions of ethyl ether and collect extracts in another separating funnel.
6. Wash ether extract with 2% aqueous KOH solution and then with water till alkali free.
7. Filter ether layer through anhydrous sodium sulphate using Whatman No 40 filter paper.
8. Evaporate the ether extract to dryness over water bath.
9. Dissolve residue in 25ml glacial acetic acid. (solution A)

(ii) Preparation of standard curve

1. Prepare 0.1% solution of pure cholesterol in glacial acetic acid.
2. Pipette aliquots of 0.1, 0.2, 0.3, 0.4, and 0.5 ml in a graduated tube and makeup the volume to 2.0ml with glacial acetic acid.
3. To each tube add 4.0ml of LB reagent and mix well.
4. Incubate the contents at 25° C for 35 minutes in water bath.
5. Quickly take optical density at 650 nm.
6. Plot a graph between concentration and optical density.

Estimation

1. Pipette 1.0ml of solution A above in a glass stopped test tube.
2. Add 4.0ml of freshly prepared LB reagent.
3. Incubate at 25° C for 35 minutes and measure O.D. at 650 nm.
4. Calculate concentration from calibration curve.
5. Report concentration of cholesterol as mg per 100g sample

   Calculation :-

   2ml solution gives concentration = A mg cholesterol

   25ml solution gives concentration = A x 25 / 2 = B mg cholesterol

   (this represent mg cholesterol in 5g sample)

   Total cholesterol (mg/100g) = Bx100/5

7. Antioxidant in ghee (Qualitative test)

Reagent : Ehrlich reagent – Mix equal volume of 0.5% sodium nitrite solution and 0.5% sulfanilic acid solution containing conc. HCl ; 72% ethyl alcohol ; 0.1N NaOH.

Procedure

1. Take 1 ml of melted fat in a test tube.
2. Add 2 ml of 72% ethyl alcohol and shake well.
3. Add 1 ml of Ehrlich reagent followed by 1 ml of 0.1N sodium hydroxide solution.
4. Observe the colour of the mixture in tube.

Result

1. Development of red-purple colour indicates presence of BHA that shows maximum absorbance at 535 nm.
2. Development of distinct salmon pink colour indicates presence of BHT that shows maximum absorbance at 505 nm.
3. Development of yellow colour, which fades rapidly, indicates presence of propyl gallate that shows no measurable maximum in visible range.

Tests for Adultrants in Ghee

Higher cost of milk fat has caught the attention of unscrupulous elements for easy adulteration with cheaper quality vegetable oils and animal fats. Many tests have been developed for checking adulteration in ghee which are based on differences in nature and contents of major and minor components of milk fat and adulterants oils / fats. To confirm purity of milk fat, generally more than one test are employed as no single test can detect all types of adulterants.

Detection of vanaspati in ghee

(a) Baudouin test

1. Take 5ml of melted fat in a test tube.
2. Add 5ml of conc. HCl and 0.2g of sucrose.
3. Mix well for one minute and keep undisturbed for 5 minutes.
4. Appearance of crimson colour indicates presence of vanaspati.

5. Add 5ml of distilled water and keep for 5 minutes.
6. Persistence of crimson colour confirms the test.

(b) Furfural test

1. Take 5ml of melted fat in a test tube.
2. Add 5ml of conc. HCl and 0.4ml of 2% furfural solution in alcohol.
3. Shake well and allow to separate the mixture, by keeping undisturbed for 5 minutes.
4. Observe colour of acid layer.
5. Appearance of pink / red colour indicates presence of vanaspati.
6. Add 5ml of distilled water, shake well and keep for 5 minutes.
7. Persistence of red colour in acid layer confirms the test.

Detection of mineral oil

1. Take 1g of clarified fat in standard joint test tube.
2. Add 5ml of 0.5N ethanolic potassium hydroxide solution.
3. Reflex the contents on boiling water bath for 10 minutes.
4. Add 5ml distilled water to saponified solution.
5. Appearance of turbidity indicates presence of mineral oil.

Detection of animal body fat

1. Take 5g of melted sample in a test tube.
2. Place in a water bath at 50 °C for 5 minutes.
3. Replace tube in another water bath maintained at 23 °C.
4. Record opacity time at regular interval using spectrophotometer.
5. Note the time taken to acquire OD of 0.14 – 0.16 at 570 nm.

Note : The opacity time for pure milk fat ranges from 12-18 minutes. Lesser time indicates presence of animal body fats. Higher time indicates presence of vegetable oils.

Detection of vegetable oils

1. Take 1-2g of molten sample and dissolve in 2-3ml of hexane.
2. Add 1.5-2.0ml of colour developing reagent (water, sulphuric acid, nitric acid in the ratio of 20:6:14).
3. Shake vigorously and stand undisturbed until it separate into two layers.
4. Appearance of distinct orange colour in upper layer indicates presence of vegetable oils.

Note : Addition of 1ml each of 25% HCl & 5% sodium nitrile solution to hexane dissolved sample, shaking well and adding 1ml of 10% sodium hydroxide solution will give orange-red colour, indicating presence of Rice Bran Oil.

Aim :- Analysis of Unsalted (cooking) butter and Table Butter

Tests are performed on table / cooking butter to have control over meeting the legal standards prescribed by Prevention of Food Adulteration Act with respect to minimum fat content and maximum moisture, curd and salt content.

Sampling : Remove several plugs from the outside to the center of the butter block and collect in a container. Mix the contents into a uniform mass. Pack it into a screw cap bottle and keep in freezer until use.

Determination of moisture

Requirements : Aluminium butter cup, tong, hot plate, dessicator, conical flask, balance, burette, etc.

Procedure

1. Take weight of empty butter cup.
2. Take 10g of well mixed sample in butter cup.
3. Heat the contents over hot plate with constant circular motion.

4. Continue heating until foaming has ceased and the curd has attained golden brown colour.
5. Allow the cup to cool in dessicator.
6. Weigh the cup.
7. Calculate moisture percentage from the differences in weights.

Calculation: Percent Moisture = B – C / B – A x 100

Where,			
	A	=	Weight of empty cup
	B	=	Weight of empty cup + sample
	C	=	Weight of cup after heating

Determination of Curd and Salt

Requirements : Aluminium dish, tong, hot plate, desiccators, balance, petroleum ether

Procedure

1. Add 10 – 15ml of petroleum ether in the above butter cup and mix well.
2. Allow to stand for settling of curd.
3. Decant the ether layer into aluminium dish.
4. Repeat extraction twice and collect ether layers by decanting.
5. Dry the cup contents over a hot plate to dryness.
6. Cool and weigh the cup.

Calculation: Percent curd + salt = D – A / B – A x 100

Where,			
	A	=	Weight of empty cup
	B	=	Weight of empty cup + sample
	D	=	Weight of cup after heating

Determination of Salt

Requirements : Conical flask, burette, beaker, water bath, potassium chromate indicator, 0.1N silver nitrate and calcium carbonate.

Procedure

1. Weigh 5g of well mixed sample, from above cup, into 250ml conical flask.
2. Add 100ml of boiling distilled water, swirl the contents and allow to stand for 10 minutes.
3. Add 2ml of potassium chromate indicator and 0.25g calcium carbonate.
4. Mix the contents and titrate against 0.1N silver nitrate solution until brick red colour persist for half minutes.
5. Carry out blank without sample.

Calculation: Percent salt = 5.85 x N x (V1 – V2) / W

Where, N = Normality of silver nitrate solution.

V1 = Volume of silver nitrate used for sample titration

V2 = Volume of silver nitrate used for blank titration

W = Weight of sample

Fat Percent = 100 – (moisture + curd + salt)

Curd percent = (curd + salt) - Salt

Aim :- Analysis of Table spreads

1. Total fat

Requirements :- Rose-Gottleib tube / Majonnier flask, fat evaporating dish, rubber stoppers, ammonia solution, diethyl ether, petroleum ether, ethyl alcohol, centrifuge machine, hot plate, oven, desiccator, weighing balance (electronic-top loading)

Procedure

1. Weigh 2-4 g of well-mixed sample in a beaker.
2. Add 3 ml of 20% HCl solution and mix well.
3. Transfer contents into extraction flask.
4. Add 10 ml of alcohol and again mix well.

5. Add 25 ml of diethyl ether, stopper the tube with cork and shake vigorously for one minute.
6. Remove the cork and add 25 ml of petroleum ether, stopper and shake well for 30 seconds.
7. Allow the tube to stand until ethereal layer is separated clearly from the aqueous layer (for Rose-Gottlieb method).
8. In case of Majonnier method- centrifuge the Majonnier flask for 10 minutes.
9. Decant the clear ethereal layer into a previously weighed aluminium fat dish.
10. Repeat extraction with 15 ml portions each of diethyl ether and petroleum ether, twice to completely extract fat from the sample.
11. Transfer the ethereal layers from each extraction into the fat dish.
12. Place the fat dish on a hot plate (60° – 65° C) and allow the ether layer to evaporate slowly.
13. Dry the residual fat at 125° C in an oven for 1 hour and then allow the fat dish to cool by placing in a desiccators.
14. Weigh the fat dish and note the weight.
15. Repeat heating, cooling and weighing until successive weighing do not differ by more than 1 mg.
16. Calculate amount of fat by difference of weight in fat dish before and after extraction (reading A).
17. Divide reading A by the weight of sample and multiply by 100 to get percent fat in sample.

    Calculation: Percent fat = (Wt. Of residue / Wt. Of sample) × 100

2. Moisture test

Requirements :- Moisture dish, Desiccators, hot air oven, weighing balance.

Procedure

1. Weigh accurately a clean and dry empty dish.(A)
2. Transfer 3-5 g of sample into the dish and note the weight. (B)
3. Place the dish on hot plate and allow the water to evaporate.
4. Place the dish in hot air oven at 105° C for 1 hours.
5. Transfer the dish to a desiccator and allow cooling for 30 minutes.
6. Weigh the contents and note the weight. (C)
7. Repeat heating and cooling until difference in two successive weights do not exceed 0.5 mg.

Observations

Weight of empty dish	=	A g
Weight of empty dish + sample	=	B g
Weight of dish after drying	=	C g

Calculations

Weight of sample	=	Reading B – Reading A
Weight after drying	=	Reading C – Reading A
Percent moisture	=	B – C / B - A x 100

3. Salt content

Requirements : Conical flask, burette, beaker, water bath, potassium chromate indicator, 0.1N silver nitrate and calcium carbonate.

Procedure

1. Weigh 5g of well mixed sample, from above cup, into 250ml conical flask.

2. Add 100ml of boiling distilled water, swirl the contents and allow to stand for 10 minutes.
3. Add 2ml of potassium chromate indicator and 0.25g calcium carbonate.
4. Mix the contents and titrate against 0.1N silver nitrate solution until brick red colour persist for half minutes.
5. Carry out blank without sample.

Calculation :- Percent salt = 5.85 x N x (V1 – V2) / W

Where, N = Normality of silver nitrate solution.

V1 = Volume of silver nitrate used for sample titration

V2 = Volume of silver nitrate used for blank titration

W = Weight of sample

4. Unsaponifiable matter and acid value of fat

When fat is saponified with excess of alkali, soap is formed and unsaponified matter is released. It consist of sterols, tocopherol and vitamin A etc. These can be extracted from saponified fat and can be estimated gravimetrically. Acidity of fat is because of the presence of free fatty acids and is an index of lypolytic rancidity, that effects shelf life of fat rich products. The estimation is based on acid-base titration and is expressed as % oleic acid.

Fat extraction

1. Weigh about 10-20g of sample (Table spread) into a beaker.
2. Add 25-30 ml of hot distilled water and mix well.
3. Transfer the contents to 250 ml separating funnel.
4. Add 25 ml of 95% ethyl alcohol and shake well.
5. Subsequently, add 60 ml of petroleum ether and 40 ml of ethyl ether. Shake the contents well.
6. Allow to stand undisturbed for few minutes.
7. Decant the solvent in a tared aluminium dish.
8. Allow the ether to evaporate at 45-50 C over a hot plate.
9. Use the extracted fat for further analysis.

(a) Determination of Unsaponifiable matter

1. Draw 5g sample of extracted fat into saponification flask.
2. Add 50 ml of alcoholic KOH and reflex for one hour.
3. Transfer the contents into 500 ml separating funnel.
4. Rinse the flask first with 100-150 ml distilled water followed rinsing with ethyl ether. Transfer all washings to separating funnel.
5. Mix the contents gently and allow to stand for 5-10 minutes.
6. Collect aqueous layer in same saponification flask and ether layer into conical flask. Make two more extractions using 2x50 ml ether portions and collect aqueous and ether layer as before.
7. Transfer the ether layer into separating funnel and wash with 3x20 ml portions of distilled water and 3x20 ml portions of aqueous solution of 0.5N KOH alternatively. Discard aqueous layer after each washing.
8. Wash excess of alkali with distilled water until wash water is alkali free.
9. Filter the contents through Whatman filter paper No. 4 with anhydrous sodium sulphate to remove traces of water.
10. Collect the filtrate in a pre-weighed flask and evaporate ether to dryness over a steam water bath or hot plate.
11. Transfer the flask in an oven maintained at 80 C for 10 minutes.
12. Cool in desiccators and weigh the residue.

    Calculation :- Unsaponifiable matter (% by weight) = (W/m) x 100

    Where W = weight of residue

    m = Weight of sample

(b) Determination of acid value

1. Weigh 5g sample of extracted fat in a 250 ml conical flask.

2. Take 25 ml of neutral alcohol in another flask and heat the contents to boiling, while still hot, neutralize the contents using 0.1N NaOH.
3. Mix the contents of two flasks and bring it to boil.
4. While still hot , titrate against standard aqueous solution of alkali (NaOH), using phenolphthalein as indicator.
5. Note the volume of NaOH used for titration.

Calculation :- Acid value = (5.61 x V) / W

Where V = Volume of standard alkali used for titration.

W = Weight of sample

(c) Phytosterol test

1. Weigh 5g sample of extracted fat in a 250 ml conical flask.
2. Add 10 ml of 1% alcoholic digitonine solution.
3. Allow the contents to cool in refrigerator for 12 hours.
4. Filter the contents through Whatman No 1 filter paper in buchner funnel.
5. Collect the precipitate of sterol digitonine and wash with cold water until foaming stops.
6. Wash with 25 ml of alcohol and diethyl ether respectively.
7. Dry the filter paper in an oven at 102° C for 10-15 minutes.
8. Collect the dry precipitate (residue) and note the weight.

Calculation :- Percent total sterol = (0.25 x P/ W) 100

Where P = Weight of precipitate(residue)

W = Weight of sample

Aim :- Analysis of milk powders including Infant milk food and weaning foods

1. Fat test

Requirements: - Rose-Gottleib tube / Majonnier flask, fat evaporating dish, rubber stoppers, ammonia solution, diethyl ether, petroleum ether, ethyl alcohol, centrifuge machine, hot plate, oven, desiccator, weighing balance (electronic-top loading)

Procedure

1. Weigh 1 - 2 g of well-mixed sample into the extraction tube.
2. Add 2-3 ml of water and 5 ml concentrated HCl and mix well by gentle swirling motion.
3. Add 25 ml of diethyl ether, stopper the tube with cork and shake vigorously for one minute.
4. Remove the cork and add 25 ml of petroleum ether, stopper and shake well for 30 seconds.
5. Allow the tube to stand until ethereal layer is separated clearly from the aqueous layer (for Rose-Gottlieb method).
6. In case of Majonnier method- centrifuge the Majonnier flask for 10 minutes.
7. Decant the clear ethereal layer into a previously weighed aluminium fat dish.
8. Repeat extraction with 15 ml portions each of diethyl ether and petroleum ether, twice to completely extract fat from the sample.
9. Transfer the ethereal layers from each extraction into the fat dish.
10. Place the fat dish on a hot plate (60° – 65° C) and allow the ether layer to evaporate slowly.
11. Dry the residual fat at 98° C in an oven and then allow the fat dish to cool by placing in a desiccators.
12. Weigh the fat dish and note the weight.
13. Repeat heating, cooling and weighing until successive weighing do not differ by more than 1 mg.
14. Calculate amount of fat by difference of weight in fat dish before and after extraction (reading A).

15. Divide reading A by the weight of sample and multiply by 100 to get percent fat in sample.

Note : For fat determination of cereal based infant food make the sample starch free by adding 0.5g diastase or amylase enzyme. Allow the starch to digest for 2 hours by keeping the extraction flask in water bath and shaking from time to time. Check by adding 2 drops of 0.1N iodine solution. No blue colour should appear.

2. Moisture and Total solid content

Requirements: - Aluminium moisture dish, boiling water bath / hot plate, weighing balance, hot air oven, desiccator, tong.

Procedure

1. Weigh accurately a clean and dry empty dish containing sand and glass rod. (A)
2. Transfer 3-5 g of sample into the dish and note the weight. (B)
3. Place the dish on hot plate and allow the water to evaporate.
4. Place the dish in hot air oven at 100° C for 3 hours.
5. Transfer the dish to a desiccator and allow cooling for 30 minutes.
6. Weigh the contents and note the weight. (C)
7. Repeat heating and cooling until difference in two successive weights do not exceed 0.5 mg.

Observations

Weight of empty dish = a g

Weight of empty dish + sample = b g

Weight of dish after drying = c g

Calculations :-

Weight of sample (A) = (Reading b – Reading a)

Weight of solids (B) = (Reading c – Reading a)

Percent Moisture (Z) = b – c / A x 100

Percent total solids = 100 – Z

3. Total Protein content.

Requirements: - Kjeldahl flask, round bottom flask, conical flask, burette, digestion heater, distillation unit, conc. sulphuric acid, copper sulphate, potassium sulphate, N/10 NaOH, 50% NaOH, N/10 sulphuric acid, mixed indicator (methyl red and 0.1% methylene blue in ethanol).

Procedure

1. Weigh accurately 0.2 g well mixed sample in a clean and dry Kjeldahl flask.
2. Add 25 ml of concentrated sulphuric acid, 0.2 g of copper sulphate and 10 g of potassium sulphate.
3. Place the flask on digestion heater under fume chamber and heat gently to boil until contents are clear, and then for another 2 hours.
4. Allow the liquid to cool and dilute with 200 ml of distilled water.
5. Fit the flask with micro-kjeldahl distillation unit.
6. Carefully add 75-80 ml of 50% NaOH down the neck of the flask to form a layer under the acid.
7. Take 50 ml of N/10 sulphuric acid in a 250 ml conical flask and place it under the condenser so that the tip of the condenser is dipped in the liquid.
8. Start heating and distil the contents until all ammonia has passed over as indicated by collection of 150 ml distillate.
9. Stop heating and remove the conical flask. Wash the condenser tip to remove all traces of condensate.
10. Titrate the contents against N/10 NaOH using mixed indicator.

11. Carry out blank determination by using 0.5 g pure sucrose in place of milk.

Observations: - Record the titre volume for original and blank experiments.

Sl.No.	Initial reading	Final reading	Vol. Of alkali used
1.			
2.			
3.			

Calculation: - Percent total nitrogen by weight = 1.4 (B-A) N / W

Percent total protein = Percent total nitrogen x 6.38

4. Total ash content

Requirements: - Single pan balance, muffle furnace, desiccator, silica crucible, tong, hot plate.

Procedure

1. Weigh the empty and dried silica crucible and note the weight.
2. Accurately weigh 1.0 – 3.0 g if sample and note the weight.
3. Evaporate the sample to dryness on a hot plate.
4. Place the crucible in a pre-heated muffle furnace and heat the contents at 550° C until ash is free from carbon.
5. Cool the crucible by placing in desiccator.
6. Weight the crucible containing the ash.
7. Continue heating, cooling and weighing until the difference in weight is not more than 0.1g.

Observations: - Record the weights.

weight of empty crucible = W g

weight of crucible with sample = W1 g

weight of crucible after drying = W2 g

weight of sample = W1 – W g (A)

weight of ash = W2 – W g (B)

Calculation: - Percent ash by weight = B / A x 100

5. Acid insoluble ash

Procedure

1. Weigh 3.0g of powder sample in a silica dish and place over hot plate till material is carbonized.
2. Remove the silica dish and place in muffle furnace at 550° C for 3 hours to obtain ash.
3. Cool and add 25ml of 5N HCl and cover with lid.
4. Heat on a water bath for 10 minutes, cool and filter through Whatman filter paper No. 42.
5. Wash with distilled water until washing is free from acid.
6. Return filter paper containing residue to the silica dish.
7. Keep contents in oven at 110° C for 3 hours followed by keeping in muffle furnace for another 3 hours.
8. Cool in desiccator and weigh.
9. Repeat process until difference between two weighings is less than 1mg.

Calculation: Acid insoluble ash % by mass(on dry matter) = 10,000 (A – B) / W (100-C)

Where, A = Weight of dish + acid insoluble ash

B = Weight of dish

C = Moisture content

W = Weight of sample

6. Total carbohydrate

This includes sucrose, maltose, dextrins, dextrose and lactose. Total carbohydrate content of infant milk food and weaning foods can be calculated by from the values obtained for moisture, fat, protein and total ash as determined earlier/ above.

Calculation - Total carbohydrate (%) = 100 – (a +b + c + d)

Where, a = Percent by mass of moisture

b = Percent by mass of total protein

c = Percent by mass of fat

d = Percent by mass of total ash

7. Iron content (Spectocolorimetric Method)

Procedure

1. Dissolve ammonium ferrous sulphate in distilled water so as to get a solution of iron concentration of 20 µg / ml. Use as standard solution.
2. Weigh 3-5g of infant food sample in a silica dish and char the sample at low heat.
3. Transfer the dish to muffle furnace at 550 C for 5 hours.
4. Cool and dissolve the ash in 5 ml of diluted hydrochloric acid (1:1) and heat gently for three minutes.
5. Cool the contents and transfer to 100 ml volumetric flask and make up the volume with distilled water.
6. Take 2-3 ml of prepared solution into measuring cell.
7. Add 3.5 ml of distilled water, 1.0 ml sulphuric acid ,1.0 ml of 7% potassium persulphate solution and 1.5 ml of 40% potassium thiocynate solution.
8. Mix well and measure optical density (OD) at 500 nm against standard iron solution.
9. Calculate iron as mg per 100g sample.

Calculation

Iron (mg/100g) = (Sc/ St.od) x (Sod/3) x (100/W) x (100/ 1000)

Where,

Sc	=	Standard Concentration
St.od	=	OD of Standard solution
Sod	=	O.D of sample
W	=	weight of sample

Reconstitution properties

1. Bulk density

The weight of powder which can be packed into a given volume is known as bulk density. It is commercially important as it effects the size of standard container, storage and transport space. For whole milk powder it varies from 0.50-0.60 g/ml and for skim milk powder the values are 0.30-0.50 g/ml.

Requirments : Weighing balance, funnel, cylinders etc.

Procedure

1. Transfer 25 g of milk powder into 100 ml measuring cylinder and note the volume.
2. Tap the cylinder 20 times with constant strokes so as allow packing of powder.
3. Again note the volume.
4. Calculate loose and packed densities of the powder sample.

   Calculation : Loose density = weight of powder / volume of powder.

   Bulk density = mass of packed material / vol. of packed material

2. Insolubility index

It is the measure of insoluble material present in the milk powder. The values for spray dried powder should not be more than 2.0 ml and for roller dried milk not more than 15 ml.

Requirements : Weighing balance, centrifuge tubes, centrifuge, beaker, stirrer.

Procedure

1. Weigh 13 g of whole milk powder or 10 g of skim milk powder into 250 ml jar.
2. Add 100 ml of distilled water at 40° C and mix the contents for 5 minutes.
3. Remove the foam and fill the liquid into 50 ml graduated centrifuge tubes.
4. Centrifuge the tubes at 3000 rpm for 5 minutes.
5. Decant the supernatant and add water upto the mark and centrifuge the contents for 5 minutes.
6. Note the volume (ml) of sediment in the tube.

3. Scorched particles

The level of scorch particles is an indicator of quality of milk as well as processing condition employed during manufacture of powder. Lower levels are required for better reconstitution properties of powders.

Requirements : Weighing balance, blending jar, cotton disc filter, mixer etc.

Procedure

1. Take 250 ml water at 40° C in the blender jar.
2. Add 25 g of skim milk powder or 32.5 g of whole milk powder and mix the contents for 60 seconds.
3. Filter through cotton disc using aspirator. Rinse the blender with 50 ml water and pass through the filter disc.
4. Remove the disc and dry it at 35°-45° C.
5. Compare the disc with standard photo print under uniform indirect light and grade the sample
6. Alternatively, weigh the disc and compare with standard print.
7. The disc showing more scorch particles than the standard disc A but less than disc B should be assigned grade B and similar for other discs.

4. Dispersibility

It is one of the properties of reconstitution of dried milks. It is affected due to changes in physical and chemical properties of milk during manufacture of powders. It is the degree of separation of wetted powder particles in water.

Requirements : Weighing balance, blending jar, brass bowl (height 13 cm, dia 24cm) fitted with an outlet, mixer, beaker, stand, standard screen etc.

Procedure

1. Clamp the bowl and stirrer onto the lab stand with clamps.
2. Take 400 ml distilled water at 40° C into the bowl.
3. Weigh 52 g powder and transfer to the surface of water.
4. Start the stirrer (400 rpm) and mix the contents for 20 seconds.
5. Release the contents through the 72 mesh screen and collect the fluid in jar and makeup the volume to 500 ml.
6. Transfer 10 ml of sample in a pre-weighed dish, in duplicate, and evaporate to dryness.
7. Cool the contents in desiccator and weigh. Calculate weigh of solids obtained.
8. Multiply by 50 to get dispersibility in grams.

5. Flowability

It is an ability of milk powder particles towards free flow to form an angle with respect to base diameter. It is expressed as angle of repose determined through the tangent which is the height of pile divided by its radius.

Requirment : Balance, glass funnel (15 cm dia) with short stem cut at a right angle, laboratory stand, plain paper sheet.

Procedure

1. Weigh 100g milk powder sample in a weighing dish.

2. Mount the glass funnel 2 cm above 40 x 40 cm plain paper sheet laid on a flat surface and pour powder in a fine stream
3. After all the powder has passed through, note the height of the cone formed.
4. Form a boundary line around the radius of the cone with a pencil.
5. Take the weight of the paper and calculate the radius from the weight per unit area of paper and the area of circular base.
6. Divide the radius of the cone by the height of the pile.
7. Express flowability as the cotangent value (cot Q) of the angle of repose determined through the tangent.
8. Flowability is directly proportional to the angle of repose.

   Cot Q = cone radius / cone hight

6. Wettability

It is the ability of milk powder particle to absorb water. During and following drying process free fat and fat migration occurs due to rupturing of fat globules that adheres onto the powder particles. This migration of fat results in formation of water repellent film in and around powder particles.

Requirement : Balance, satin cloth piece, hollow tin pipes (6.5 cm high and 5.0 cm in dia), stop watch, shallow water dish.

Procedure

1. Weigh accurately 2.5g of powder sample in a weighing dish.
2. Secure the satin cloth onto a hollow tin pipe with the help of rubber band.
3. Place the hollow pipe into a shallow water dish containing water at 40° C so that the cloth touches the water surface.
4. Spread weighed amount of sample over the cloth evenly.
5. Immediately start the stop watch and note the time for the powder to get wet completely.

6. Take mean of three tests as wetting time in seconds.

7. Sinkability

It is the ability of powder particles to completely wet and sink in the water. It is expressed as milligram of powder sinking per minute per square centimeter.

Requirement : Balance, pipette, spectrometer with chart recorder, cuvettes, stop watch, butter paper.

Procedure

1. Weigh accurately 10 mg powder sample on a butter paper.
2. Measure 3.5 ml of distilled water in a pipette and transfer into the spectrometer cuvette.
3. Adjust the measuring wave length to 760 nm.
4. Place the cuvette in spectrometer and start the recorder.
5. Dust the powder sample on the surface of distilled water in the cuvette.
6. Start the stop watch immediately and note the percent transmission for a period of 6 minutes at an interval of 2 minutes.
7. The curve thus obtained signifies the sinkability characteristics of the sample.

8. Free fat content

During and following spray or roller drying process free fat and fat migration occurs due to rupturing of fat globules that adheres in and around the powder particles. Excess of free fat results in oxidative deterioration especially in whole milk powder.

Requirement : Weighing balance, conical flask (250 ml), filter paper (No. 42) aluminium dish, hot plate, petroleum ether.

Procedure

1. Weigh 10g of powder sample in a conical flask.

2. Add 100 ml of petroleum ether and shake the contents vigorously in a vertical motion ten times.
3. Allow to settle for 15 minutes and filter.
4. Collect the filtrate in a pre-weighed aluminium dish.
5. Repeat extraction once more and collect the filtrate.
6. Evaporate the solvent on a hot plate.
7. Cool and weigh the dish containing the fat.
8. Express result as percent free fat.

9. Lactate content

Requirements : Conical flask, pipette, volumetric flask, test tubes, glass stopper, balance, spectrophotometer/ colorimeter, water bath.

Reagents

1. Copper sulphate solution - Dissolve 250g of copper sulphate,volumetrically in 1000ml distilled water.
2. Calcium hydroxide suspension – Grind 300g calcium hydroxide in mortar with 900ml distilled water and keep in tight stoppered bottle.
3. Copper sulphate – sulphuric acid solution – Add 0.5ml of copper sulphate solution to 300ml of sulphuric acid.
4. Hydroxydiphenyl solution – Dissolve 0.75g p-hydroxydiphenyl in 5ml aqueous solution of NaOH (50g/L) and dilute to 50ml with distilled water. Keep in coloured bottle in cool dark place.
5. Lithium lactate standard solution – Dissolve 0.1067g of lithium lactate in 1000ml distilled water. This corresponds to 0.1mg lactic acid per ml. (prepare fresh).

Sample preparation: Calculate percent by mass "a" of SNF in the sample, using the formula "a" = 100 – (F – M) where, F is fat content of sample and M is the moisture content of sample.

Weigh 0.1g "a" mass of test sample = 1000g / a-100. Add this portion in 100ml distilled water and mix well.

Preparation of standard curve

1. Pipette 5ml of reconstituted skim milk having 30mg lactic acid / 100g SNF into each 50ml volumetric flask.
2. Add 0.0, 1.0, 2.0, 3.0, and 4.0 ml of lithium lactate solution to obtain a range of standard corresponding to 0, 20, 40, 60 and 80 mg added lactate/100g SNF.
3. Dilute each standard with 35 ml of distilled water.
4. To each tube add 5.0ml of copper sulphate solution followed by 5ml of calcium hydroxide solution mix well and allow to stand for 10 minutes.
5. Dilute upto mark with distilled water, allow to stand for 10 minutes, filter through Whatman No.1 filter paper and collect filterate.
6. To 1.0ml of filtrate add 6.0ml copper sulphate-sulphuric acid solution and heat for 5 minutes over boiling water bath.
7. Add 2 drops of p-hydroxydiphenyl solution, shake well and keep at 30 C for 15 minutes and then on boiling water bath for 90 seconds.
8. Cool and measure absorbance at 570 nm.
9 Plot a graph between absorbance and masses of lactic acid to obtain calibration curve.

Estimation

1. Pipette 5.0ml of sample solution into 50ml volumetric flask and dilute with 35ml of distilled water.
2. To each tube add 5.0ml of copper sulphate solution followed by 5ml of calcium hydroxide solution mix well and allow to stand for 10 minutes.
3. Dilute upto mark with distilled water, allow to stand for 10 minutes, filter through Whatman No.1 filter paper and collect filtrate.

4. To 1.0ml of filtrate add 6.0ml copper sulphate-sulphuric acid solution and heat for 5 minutes over boiling water bath.
5. Add 2 drops of p-hydroxydiphenyl solution, shake well and keep at 30° C for 15 minutes and then on boiling water bath for 90 seconds.
6. Cool and measure absorbance at 570 nm against blank.
7. Prepare a blank using 35ml distilled water and treating it in same manner as for sample.
8. Calculate mg lactic acid / 100g SNF by reference to calibration curve.

10. Particle size

Procedure

1. Weigh 100g of powder sample in 150 micron sieve and cover it.
2. Place it over 75 micron sieve with a receiver container.
3. Place the sieve nest on a sieve shaker and shake for 15 minutes.
4. Transfer the residue of each sieve, separately, to two tarred containers.
5. Calculate percent material retained on each sieve.

Calculation :-

A) Material retained on 75 micron sieve (%mass) = (A+B / W) x 100

B) Material retained on 150 micron sieve (%mass) = (B / W) x 100

Where, A = weight of material retained on 75 micron sieve

B = weight of material retained on 150 micron sieve

W = Weight of test material

Aim :- Analysis of Edible Casien

Sample preparation :- Take 50g of sample and grind it to fine powder. Pass the sample through a 500 μm sieve. Use sieved material for further testing.

1. **Moisture content** – Similar to the method described for cheese.
2. **Nitrogen content** - Similar to the method described for cheese.
3. **Total Ash content** - Similar to the method described for cheese.
4. **Fat content** – Take 5 g of sample in 20 ml conc. HCl, heat and boil for 10 minutes. Transfer quantitatively to mojonnier flask and proceed as in case of cheese.

11. Total acidity

Procedure

1. Take 2g of sample into 250 ml flask and add 10 ml distilled water.
2. Soak for 15 minutes, add 50 ml of 0.1N sodium hydroxide solution and heat at 50 C in water bath.
3. Titrate with 0.1N sulphuric acid solution using 0.04% bromothymol blue as indicator.
4. Make a blank determination without casein.
5. Express acidity as ml of 0.1N NaOH required to neutralize 1.0g sample.

12. Free acidity

Procedure

1. Take 10g casein in 500 ml conical flask containing 200 ml hot distilled water.
2. Allow to stand for 30 minutes with frequent agitation.

3. Filter through Whatman filter paper No.1 in 250 ml conical flask.
4. Pipette 100 ml, of cold filtrate (corresponding to 5g casein) into a beaker.
5. Add 0.5 ml of alcoholic phenolphthalein and titrate against 0.1N sodium hydroxide solution.
6. Express free acidity as ml of 0.1N NaOH required to neutralize 10g sample.

Calculation :- Free acidity = (20 x V x N) / W

Where V = volume of standard NaOH solution

N = Normality of NaOH solution

W = weight of sample

13. Acid insoluble ash

Procedure

1. To the ash in crucible add 25 ml of 5N HCl and heat for 5 minutes over water bath.
2. Cool and filter through Whatman No. 42 filter paper.
3. Wash filter paper 5-6 times with water until washings are acid free.
4. Place filter paper in crucible and keep inj oven at 120-130 C for 30 minutes.
5. Ignite in muffle furnace at 550 C for one hour.
6. Cool and weigh the dish until constant weight.

Calculation : Acid insoluble ash (%by weight) on dry basis = 100(B-A)/ C-A

Where B = weight of crucible with ash

A = weight of empty dish

C = weight of dish + dry matter taken for ash determination

14. Lactose content

Preparation of standard curve

1. Pipette 10 ml of lactose solution (20g lactose monohydrate/ 100 ml distilled water) into 100 ml volumetric flask and dilute with distilled water.
2. Pipette 1,2,and 3 ml of above solution in 100 ml volumetric flask and make up the volume with water.
3. Take 2 ml of standard solution in 40 ml of glass stoppered test tube.
4. Add 0.2 ml of 80% phenol solution using micropipette and mix well.
5. Add 5 ml of concentrated sulphuric acid and mix well.
6. Allow to stand for 15 minutes, cool for 5 minutes in ice-bath.
7. Immediately measure absorbance at 490 nm against reagent blank.
8. Draw a calibration graph by plotting absorbance against lactose concentration in mg/ml.

Blank solution :- weigh 0.1g of sodium hydrogen carbonate, add 15 ml of water, 8 ml of concentrated HCl and 1.0 ml of 1% acetic acid solution. Mix the contents and allow to stand for 5 minutes. Add 1.0 ml of 0.1N sodium acetate solution, filter and follow steps "c – g" above.

Procedure

1. Take 0.1 g of test sample, add 25 ml of water and place in water bath at 65 -70 ° C with occasional mixing.
2. Cool and add 15 ml of water followed by addition of 8 ml concentrated HCl and 1.0 ml of 1% acetic acid solution.
3. Stopper and mix the contents by shaking and allow to stand for 5 minutes.
4. Add 1.0 ml of 0.1N sodium acetate solution and mix well.

5. Allow the casein to precipitate and filter. Collect filtrate. Discard first 10 ml of filtrate.
6. Take 2 ml of clear filtrate in 40 ml of glass stoppered test tube.
7. Add 0.2 ml of 80% phenol solution using micropipette and mix well.
8. Add 5 ml of concentrated sulphuric acid and mix well.
9. Allow to stand for 15 minutes, cool for 5 minutes in ice-bath.
10. Immediately measure absorbance at 490 nm against reagent blank.

Calculation :- Lactose content (%) = (c/106 x 50 / m) × 100

Where c = concentration of lactose(from calibration curve)

m = weigh of test sample

❑❑❑

Chapter–6

Tests to Maintain Quality of Dairy Products

The growing chemicalization of milk presents a great challenge to the dairy industry. A number of chemical compounds get entry into milk, intentionally or un-intentionally, through their direct or indirect usage for crop, animal; or dairy husbandry practices, as environmental contamination. Several of these contaminants may be naturally present on grazing grass or use as pesticides for green fodder or they may get entry through use of compounded animal feeds. These eventually get excreted into milk as residues and pose problems to food safety. The risk of health hazards arises from veterinary drugs, residual pesticides, heavy metals, detergents and micotoxins *etc.* It is therefore, become necessary to detect these contaminants in order to make food safe for human consumption.

Detection of Aflatoxin M 1 in Milk

Material requirements : Glass column(30 cm x 1 cm), Silica gel (63-200 μg particle size), volumetric flasks, beakers, micro pipette, HPLC equipment fitted with C 18 column(25 cm x 4.6 mm) and fluorescence detector.

Reagents: Sodium chloride solution (400g in 1000ml water), acetonitrile/water mixture (25:75), sodium sulphsate, chloroform, toluene/acetic acid mixture (9:1), acetonitrile/diethyle ether/hexane mixture (1:3:6), chloroform/acetone mixture (8:2), n-hexane and aflatoxin standard ampule.

Preparation of stock solution : Transfer contents from 2.5ml or 0.5ml ampoule, quantitatively, to 25ml or 50ml volumetric flask respectively. Rinse ampoule with small quantities of chloroform and collect rinsing into flask. Makeup the volume with chloroform, mix well and store at 5° C or below. These solutions will contain 1 µg per ml aflatoxin M1.

Working solution : Transfer 0.5ml of stock solution to a 10ml conical flask. Evaporate contents to dryness using inert gas. Dissolve residue in 10ml of acetonitrile/water mixture. Prepare series of dilutions containing 0.0025, 0.005, 0.01 and 0.025 µg/ml aflatoxin. Similarly treat 5ml of stock solution to give a concentration of 0.5 µg/ml.

Extraction

1. Fill, glass column fitted with sintered disc at the bottom, with 10ml chloroform.
2. Add 1g sodium sulphate and rinse with 1ml chloroform to remove particles and to ensure that sodium sulphate layer is horizontal.
3. Transfer 2g silica gel into column and add 5ml chloroform, stir the mix until silica gel is dispersed.
4. Drain off excess chloroform leaving 3 cm above gel layer.
5. Cover gel with 2g sodium sulphate and cover with chloroform.
6. Transfer 75ml of filtrate and allow to elute until level of liquid reaches top surface of sodium sulphate layer.
7. Add 25ml of toluene/acetic acid mixture and elute the same.
8. Now elute with 25ml n-hexane followed by 25ml acetonitrile/ diethyle ether/ hexane mixture.
9. Finally elute aflatoxin with 60ml chloroform/acetone mixture and collect in round bottom flask.

10. Evaporate at 50° C under inert gas, using rotary evaporator, until dry.
11. Transfer residue using chloroform to a 4ml conical tube.
12. Cool and add 100 µl acetonitrile/water mixture and mix using vortex mixer for one minute.

Determination

1. Inject 2,4,6,8 and 10 µl of 0.05 µg/ml aflatoxin solution (equivalent to 1,2,3,4 and 5 µg/ml) into HPLC column.
2. Elute using acetonitrile/water mixture as a mobile phase, at a flow rate of 1ml/minute.
3. Aflatoxin is detected using florescence detector at 360 and 420 nm.
4. Obtain standard chromatogram
5. Alternatively, prepare a calibration curve by plotting peak area of standards as a function of quantity of aflatoxin injected.
6. Inject 40 µl of sample extract into HPLC column and elute using similar conditions.
7. Determine area of aflatoxin peak of sample extracted.
8. Determine amount of aflatoxin in µg from calibration curve.
9. If peak area of sample is greater than standard, dilute extract, quantitatively, with mobile phase and inject again and follow step 6 &7 to know amount of aflatoxin in sample.
10. Express aflatoxin as µg per liter of sample

    Calculation : Aflatoxin (µg/L) = (Maf x Vext) / (Vm x V) (Vf / 125)

    Where,

    Maf = mass (ng) of aflatoxin determined from calibration curve.

    Vext = Volume (µl) in which the sample extract is dissolved.

Vm = Volume (μl) of sample extract injected.

V = Volume (ml) of milk.

Vf = Volume (ml) of filtrate obtained from extracted stage

125 = Volume (ml) of chloroform used for extraction.

Note : The ratio of two single results on identical sample should not exceed 1.9. The ratio between independent results obtained by two operator at different laboratory on identical sample should not exceed 3.2.

Determination of Malamine in Milk

Malamine is a trimer of cyanamide and contains 66% nitrogen by mass. It is added to milk to increase apparent nitrogen content resulting in higher protein content than true value. When added to milk it leads to kidney stones and other renal failure, especially in infants and young children. It has recently come to news after several companies were implicated of its presence in their milk and milk products. A reliable, rapid and inexpensive method of detection is described hereunder.

Requirements :- Glass plates 10x10cm, TLC chamber, sample applicator, Vertical chamber, Densitometer.

Potassium permanganate solution – Dissolve 8g of potassium permanganate in 100ml distilled water.

Hydrochloric acid solution – Mix 25ml of 32% hydrochloric acid in 50ml distilled water.

Wusters reagent– Dissolve 500mg of N,N,N2 ,N2 ,tetramethyl-1-4-phenylammonium chloride in 100ml acetone.

Starch-iodine solution -

(i) Dissolve 800mg of potassium iodide in 20ml distilled water.

(ii) Dissolve 800mg of starch in 20ml distilled water.

Mix the two solutions and dissolve in 10ml ethanol.

Developing reagent (mobile phase) – Mix 2-proponol, CH_2Cl_2 and water in the ratio of 3:1:1

Test 1. (TLC METHOD)

Requirements :- Glass plates 10x10cm, TLC chamber, sample applicator, Vertical chamber, Densitometer.

Potassium permanganate solution – Dissolve 8g of potassium permanganate in 100ml distilled water.

Hydrochloric acid solution – Mix 25ml of 32% hydrochloric acid in 50ml distilled water.

Wusters reagent– Dissolve 500mg of N,N,N2 ,N2 ,tetramethyl-1-4-phenylammonium chloride in 100ml acetone.

Starch-iodine solution -

(i) Dissolve 800mg of potassium iodide in 20ml distilled water.

(ii) Dissolve 800mg of starch in 20ml distilled water.

Mix the two solutions and dissolve in 10ml ethanol.

Developing reagent (mobile phase) – Mix 2-proponol, CH_2Cl_2 and water in the ratio of 3:1:1

Procedure

1. Coat the glass plates with K60 Silica gel and allow to dry.
2. Apply 1 μl volume of sample as 7mm bands using Degas sample applicator.
3. Develop the plates in TLC chamber containing the mobile phase (without chamber saturation) for 35 minutes or to distance of 55mm.
4. Dry the plates for 5 minutes in gentle stream of air.
5. Place the plates in a vertical chamber containing potassium-permanganate- HCl solution for 5 minutes.
6. Dip the chlorinated plates in Wusters reagent for 2 seconds to form deep blue zones.
7. Scan the plates at 550-615 nm wavelength using Diode-Array scanner within six hours of dipping in Wusters reagent.

8. When starch-iodine reagent is used for developing zones, the chlorinated plates are dipped for 1 second and scanned at a wavelength of 490-610 nm.
9. Measure peak appearance distance and corresponding Rf values.

Note : - Melamine peak is expected to appear at a distance of 18.7 mm with an Rf value of 0.30.

Test 2. (Elisa Method)

Requirements :- Melamine ELISA kit consisting of - microfilter plates (12 x 8 strips) coated with melamine antibody, standards solutions (0, 20, 50, 100, 200, 500 ppb), melamine- enzyme conjugate, substrate solution(TMB), dilute HCl solution, wash buffer(5X concentrate) and deionized water, multichannel pipette or stepping pipette, ELISA photometer.

Dilute wash buffer in ratio of 1:5 with deionized water before use.

Adjust microplate and reagents at room temperature before use.

Procedure

1. Transfer 100 µl of the standard solution or sample extract into microplate wells in duplicate.
2. Using multichannel pipette, add 50 µl of enzyme conjugate solution in each wells.
3. Cover the wells with parafilm and mix the contents with circular motion for 30 seconds.
4. Incubate the strip for 30 minutes at room temperature.
5. Wash the strip for four using 250 µl of diluted wash buffer solution for each washing.
6. Add 100 µl of substrate solutionto the wells and incubate for 20 minutes at room temperature. Do not expose strip to direct sunlight.

7. Add 100 µl of dilute HCl solution to wells and incubate for 20 minutes at room temperature.
8. Read absorbance at 450 nm using ELISA photometer, within 15 minutes of addition of dilute HCl solution.
9. Calculate the mean absorbance value of each of the standards and draw a standard curve between absorbance value and concentration (ppb).
10. Determine concentration using standard curve.
11. Alternatevily, compare sample absorbance with absorbance of calibrator.

Observation : Sample containg less colour than calibrator has graeter concentration than calibrator and visa-versa.

Detection of Lead in Milk

Standard solution – Dissolve 1.5985g purified lead nitrate (99%) in 1000ml of 1N nitric acid. This solution contains 1mg/ml or 1000 ppm of lead.

Intermediate solution – Pipette 10ml of stock solution into 1000ml volumetric flask and makeup the volume with nitric acid. This solution contains 100 ppm of lead.

Working solution – Pipette 2,4,6,8 and 10ml of 100 ppm solution into 100ml volumetric flask and dilute with 1n nitric acid. These dilutions corresponds to 2,4,6,8, and 10 ppm of lead

Sample preparation

1. Weigh the empty and dried silica crucible and note the weight.
2. Transfer 50 ml of milk sample and note the weight.
3. Evaporate the sample to dryness on a hot plate.
4. Place the crucible in a pre-heated muffle furnace and heat the contents at 550° C until ash is free from carbon.
5. Allow to cool over night.
6. Dissolve residue (white ash) in 5ml of 1n nitric acid by heating slowly over a hot plate.

7. Filter using ashless filter paper into 25ml volumetric flask.
8. Rinse contents of crucible with small quantities of 1N nitric acid, filter and collect rinsing in flask.
9. Prepare a blank without sample.

Determination

1. Optimize burner position by focusing cathode lamp on target area of cleaning strip.
2. Adjust lamp position to get maximum intensity at 217 nm.
3. Adjust slit width to 1.0 mm.
4. Ignite flame and aspirate standard solution to get calibration curve.
5. Similarly, aspirate blank followed by sample to record concentration of lead in sample.

   Calculation :- Pb (ppm) = (sample reading – blank reading) x final dilution wt. Of sample

Determining Strength of Washing Solution

Alkalinity test

The strength of washing solution is generally expressed in terms of NaOH as it is often used as such or in combination with disodium phosphate and sodium metasilicate etc. An alkali equivalent to 1.5% NaOH is generally used for glass bottles and 0.5% for can washing. Repeated usage of washing solutions in automatic machines generally gets diluted. It is important to maintain the proper alkalinity of detergent solution for efficient cleaning and germicidal effects.

Qualitative test :- Take 10 ml of detergent solution in conical flask. Add 12.5 ml of n / 10 HCL and 5 drops of 0.5 % phenolphthalein indicator. If the mixture turns red it indicates alkalinity greater than 0.5 % as caustic soda, which is sufficient.

Quantitative test. (**Direct reading method) :-** For quick determination for daily use.

Procedure

1. Measure 10 ml of sample in 100 ml conical flask
2. Add 5 drops of phenolphthalein indicator.
3. Titrate with 2.5 N sulphuric acid until solution becomes colourless.
4. Note milliliters of acid used for titration.
5. The acid used is the direct indicator of percent alkali as NaOH.
6. Run the test in duplicate and take average of two titer readings for calculations.

Laboratory method

Procedure

1. Take 5 ml of washing solution in 250 ml volumetric flask.
2. Makeup the volume with distilled water and mix well.
3. Transfer 50 ml of this solution in 100 ml conical flask.
4. Add few drops of phenolphthalein indicator.
5. Titrate the contents with 0.1 N sulphuric acid to colourless solution.
6. Record ml of acid used as "A".
7. Add few drops of methyl orange indicator.
8. Continue titration until slight pink colour appears.
9. Note ml of acid used as "B".

   Calculation :- % free caustic = (reading A – reading B) x 0.4 % total alkali = (A + B) x 0.4

Acidity of cleaning solution

Requirements: Burette, pipette, conical flask, beaker and 0.1N sodium hydroxide solution, Phenolphthalein indicator.

Procedure

1. Take 10ml sample of cleaning solution in 250ml conical flask.
2. Add five drops of phenolphthalein indicator.
3. Titrate with 0.1N sodium hydroxide solution until appearance of pink colour.

   Calculation :- Acidity expressed as HCl acid = volume of 0.1N NaOH used x 0.0365

Determination of available chlorine in sodium hypochloride solution

As the strength of chlorine solution decreases on storing it is necessary to check its strength regularly. The strength of stock solution of chlorine is about 35% in terms of available chlorine. The method to determine available chlorine in solution is based on reaction between available chlorine from hypochloride solution and acidified potassium iodide solution in which iodine is librated. The librated iodine is titrated against standard sodium thiosulphate solution using starch as indicator. The quantity of available chlorine can be calculated from number of milliliter of sodium thiosulphate used. Available chlorine which is librated from hypochloride solution, by the addition of acid, librates equivalent amount of iodine from potassium iodide which produces yellowish brown colour.

Stock solution :- Weigh 10 g of bleaching powder and make into paste with small quantity of water in pastel and mortar. Transfer the contents, quantitatively, to 250 ml volumetric flask using distilled water. Make up the volume and mix well.

Procedure

1. Pipette 10 ml of sodium hypochloride solution into 250 ml volumetric flask and makeup the volume with distilled water.
2. Pipette 25 ml of diluted solution in a 250 ml conical flask.
3. Add 2 g potassium iodide crystal followed by 10 ml glacial acetic acid.

4. Titrate the contents with N/10 sodium thiosulphate until brown colour changes to light straw yellow.
5. Add 1 ml of 1% starch solution and continue titration till colour disappears.
6. Record volume of thiosulphate used as "V" ml.
7. Make a blank determination using same reagents and deduct the titer value from "V" ml.
8. Record values of both titrations.

   Calculation :- 1 ml of N/10 Sod.thiosulphate = 0.003546 g available chlorine

   V ml of N/10 Sod.thiosulphate = V x 0.003546 (chlorine from 25 ml solu.)

   Chlorine in 250 ml solu. = V x 0.003546 x 250 / 25 (chlorine in 10 ml stock solu.)

   Percent Chlorine (%) = V x 0.003546 x 250/25 x 100/10

Estimation of Iodine concentration in chemical sanitizers

Iodine is an antimicrobial agent used for sanitization of dairy equipments along with surfactant as a carriers and are called "IDOPHORES". Most effective agent is the dissociated free chlorine. The degree of effectiveness depends on properties of surfactants used in formulation. Idophores, like chlorine compounds, are active against bacteria, viruses, yeasts and moulds. It is an ideal cold sanitizer. The bactericidal activity of iodine is through direct halogenation of proteins resulting in cell wall damage and destruction of microbial enzyme activity.

Requirements: Burette, 250ml conical flask, beaker, pipette and 0.01N sodium thiosulphate solution.

Procedure

1. Pipette 100ml of sanitizer test solution in a 250ml conical flask.

2. Fill burette with 0.01N sodium thiosulphate solution.
3. Titrate the contents of conical flask with 0.01N sodium thiosulphate solution till disappearance of yellow colour .

   Calculation: 1ml of 0.01N sodium thiosulphate = 12.7 ppm iodine Iodine concentration (ppm) = Volume of sodium thiosulphate used x 12.7

Assessing sterility of dairy equipments

Cleaning and sanitization is an integral part of dairy operations. Improper cleaning and sterilization of dairy equipments and utensils may contribute towards contamination of milk and milk products. It is therefore, necessary to test dairy equipments periodically for sterility. The sanitary conditions of milk cans, pails and other containers may be tested by rinse test while storage tanks, vats, pipelines, heat exchangers, and other large equipments may be tested using swab test.

Requirements :- Dilution water, agar media, Ringers solution, Conical flask, Pipettes, Water bath, Incubator, Petri dishes, test tubes, dilution bottles, wire gauge, cotton wool etc.

Rinse test

Procedure

1. Add 500 ml of Ringers solution in the container (milk can) to be tested.
2. Rotate the container gently for 12 times in each direction so as to wet whole of internal surface and transfer the sample into sterile flask.
3. Transfer 1.0 ml portion of the rinse sample in petri dish, in duplicate, pour the pre-melted agar media and allow to set.
4. Incubate the petri plates at 37° C for 48 hours and count the number of colonies formed.
5. Take average count in duplicate plate and multiplied by 20.
6. Express result in terms of counts per can, which is colony count per ml multiplied by 500.

7. Colony count per litre capacity = count per can / capacity of can.
8. Grade the sanitary conditions of milk cans as follows

Counts / Lit. capacity	**Sterility**
Less than 1000	Satisfactory
1000 – 5000 1000 – 5000	Fair
More than 5000	Unsatis

Swab test

Procedure

Swab preparation : Take a piece of metal wire 35 cm x 2.6 mm formed into a loop at one end. Wind non-absorbent cotton wool around the notched end over a length of 50 mm and secure the swab with thread.

1. Place the swab in a test tube containing Ringers solution and plug the tube mouth with cotton plug. Autoclave the contents at 121° C for 15 minutes.
2. Press the swab with rolling motion against tube wall to remove excess liquid.
3. Rub the swab, back and forth with pressure and rotating action, over the surface area to be examined. Cover several 10 x 5 cm area from different parts of the surface. Total area to cover under swabing-900 sq.cm.
4. Return the swab in original liquid and immerse for 5 minutes and then mix by swirling the swab vigorously in the solution.
5. Remove swab after draining excess liquid by pressing against tube wall. Mix the solution thoroughly.
6. Transfer 1.0 ml of the well mixed swab sample in duplicate petri plates, pour agar media and allow to set.
7. Also prepare similar plates using 1/10 dilution of the sample.
8. Incubate the petri plates at 37° C for 48 hours and count the colonies in each plate and find colony count per ml of swab sample.

9. Match the counts with following standard as a guide for judging sanitary conditions.

   Calculation: Colony count / 900 sq.cm. = colony count / ml x 25x 900 / area swabbed

Number of colonies per 900 sq.cm. area	Sterility
Less than 5000	Satisfactory
5000 – 25,000	Fair
More than 25,000	Unsatisfactory

Test for water hardness

Hardness is imparted to all fresh water by alkaline metal cations of calcium and magnesium in combination with carbonates and bi-carbonates, sulphates, chlorides and nitrates. Hard water causes problems of scaling, corrosion, reduce cleaning efficiency of detergents. For good portable water the hardness should be below 20mg/L.

Principle:- Compounds of calcium and magnesium ions form wine red complex with Erichrome T. The di-sodium salt of EDTA extracts the metal ions from dye metal ions complex as colorless complexes.

Requirements :- Conical flask, burette, stand, beakers, ammonium buffer solution, Erichome black T indicator, Di-sodium EDTA salt (0.01M solution).

Procedure

1. Draw 50 ml water sample in a conical flask.
2. Add 1ml of ammonium buffer solution and 5 drops of indicator solution (Erichrome black T – 0.05%).
3. Titrate with 0.01 M EDTA solution until clear blue color appears and note the reading.
4. Repeat titration and take the average value of titer.
5. Express result as mg/L as calcium carbonate.

Calculation :- Total hardness = ml of EDTA used x 1000 / vol. of sample

Proficiency testing

Analytical Quality Assurance and Good Laboratory Practice (GLP) in dairy laboratories are major issues that are constantly gaining importance especially in context to GATT negotiations of Uruguay round. Milk and dairy products are highly perishable products, which can not be traded over long distances without a great deal of care. Therefore it becomes obligatory to develop standards and specifications for these products to build up international confidence about the reliability of their composition. Such International Confidence can only be achieved if the participating countries in trade can rest assured the consumers that the goods offered by the manufacturing countries are subjected to quality and safety control in accordance with the rules of good manufacturing practices.

Proficiency testing is the use of results generated in inter laboratory test comparison for the purpose of assessing the technical competence of participating testing laboratories. The scheme provides laboratories with an objective means of assessing and documenting the reliability of data they are producing. The common feature of proficiency testing scheme is that the test results obtained by one laboratory are compared with those obtained by one or more testing laboratories. The laboratories participating in proficiency testing schemes must use method of analysis which are validated as fit for the purpose before use and fully documented.

The proficiency testing schemes must provide a transparent interpretation and assessment of results. Laboratory wishing to demonstrate their proficiency should seek and participate in proficiency testing schemes relevant to their area of work. Bodies assessing the technical competence of testing laboratories such as accreditation bodies use the results of proficiency testing in their assessment of technical competence to varying extent in their evaluation procedures. The proficiency testing is the use of inter laboratory test comparison for checking over all laboratory performance.

The purpose for which inter-laboratory test comparisons are undertaken may include:

- Checking overall laboratory testing performance
- Checking Individual testing performance of laboratory staff
- Establishing that effectiveness of a test method
- Determination of one or more characteristics of a material or product to a particular degree of accuracy.

Various protocols for proficiency testing schemes have been produced. The need now is for a harmonized protocol that will be universally accepted. Proficiency testing methods vary depending on the nature of material or product under test, the test method in use and the number of testing laboratories participating. They possess the common feature of comparison of test results obtained by one testing laboratory with those obtained by one or more other testing laboratories.

Testing of dairy effluents

1. Biological oxygen demand (BOD)

Principle: Bio-Chemical (or Biological) Oxygen Demand is a measure of the degradable organic matter present in water sample and can be defined as the amount of oxygen required by the microorganisms for stabilizing the biologically degradable organic matter under aerobic conditions.

Requirements : B.O.D bottles 300ml capacity,B.O.D. incubator $20^0C \pm 1^0C$

Phosphate buffer : Dissolve 8.5gm Potassium Dihydrogen phosphate, 21.75gm Dipottasium hydrogen phosphate, 33.4gm Disodium hydrogen phosphate heptahydrate and 1.7gm ammonium chloride in 500ml distilled water and dilute to 1 litre. It should have a pH of 7.2 without adjustment.

Magnesium sulphate : Dissolve 22.5gm Magnesium Sulphate in distilled water and dilute to 1 liter.

Calcium chloride: Dissolve 27.5gm Anhydrous Calcium Chloride in distilled water and dilute to 1 liter.

Ferric chloride: Dissolve 0.25gm Ferric Chloride ($FeCl_3.6H_2O$) in distilled water and dilute to 1 liter.

Manganese sulphate **:** Dissolve 480gm Manganese Sulphate ($MnSO_44H_2O$) or 400gm Manganese Sulphate ($MnSO_42H_2O$) or 364gm Manganese Sulphate ($MnSO_4H_2O$) in distilled water, filter and dilute to 1liter. It should not give a color with starch when added to acidified potassium iodide solution.

Alkali-azide iodide reagent: Dissolve 500gm of Sodium Hydroxide or 700gm Potassium Hydroxide and 135gm Sodium iodide or 150gm Potassium Iodide in distilled water and dilute to 1 litre.To this solution add 10gm Sodium azide dissolved in 40 ml water.

Concentrated sulphuric Acid: 1ml of Concentrated Sulphuric Acid corresponds to 3ml of Alkali Azide Iodide reagent.

Sodium thiosulphate stock solution, 0.1N: Dissolve 24.82gm Sodium thiosulphate in boiled and cooled distilled water and dilute to 1 liter.

Starch solution: Add 5gm starch to approximately 800ml of boiling water with stirring. Dilute to 1litre, boil for few minutes and allow to stand overnight. Use clear supernatant.

Standard sodium thiosulphate, 0.025N**:** Prepare by diluting the stock Sodium thiosulphate Solution.

Procedure

1. In a stainless steel vessel of 5litre capacity add 2 liter glass distilled water which has been previously saturated with oxygen overnight. Add 2.0ml each of Phosphate Buffer, Magnesium Sulphate Solution, Calcium chloride Solution and Ferric Chloride Solution. This is Dilution Water.
2. Neutralize the effluent /wastewater sample to pH 7.0, store the neutralized sample in open for 1-2 hours before use to remove residual chlorine.
3. Take suitable proportion of the effluent sample (5%—25%) in 1litre volumetric flask and dilute with dilution water to 1 liter. Mix thoroughly.

4. Fill 2 BOD bottles (300ml each) with above solution and stopper, excluding air bubbles. Keep one bottle for 5days in BOD Incubator at $20^0C \pm 1^0C$ and determine the dissolved oxygen after 5 days.

5. To the another bottle , add 2ml Manganese Sulphate Solution and add 2ml Alkali Azide Iodide Reagent, well below the surface of the liquid, stopper excluding air bubbles and mix by inverting at least 15 times. After Manganese Sulphate precipitate settles, shake again and allow the precipitate to resettle.

6. After 2-5 minutes, while at least 100ml clear supernatant is there, add 2ml concentrated Sulphuric Acid by the side of the bottle and restopper. Mix to dissolve by gentle inversion. This has Iodine which corresponds to 200ml of the original sample, after correction for the loss of the sample by displacement with the reagents has been made. Thus, when a total of 4ml (2ml Manganese Sulphate solution and 2ml Alkali Azide Reagent) are added to a 300ml BOD bottle the volume taken for titration should be 200*300/300-4 = 203ml.

7. Titrate 203ml with 0.025N Standard Sodium Thiosulphate Solution to a pale straw color. Add 1-2ml Starch solution and titrate to disappearance of blue color.

   a. 1ml of 0.025N Sodium Thiosulphate Solution ae 0.200mg Dissolved Oxygen

   b. 1ml of Titrate ae 1mg/litre of Dissolved Oxygen

8. Prepare blank dilution water sample and repeat the steps as above.

9. At the end of incubation, titrate as per step-5.

Calculation :-

$$\text{B.O.D mg/ liter} = \frac{(D1 - D5) - (B1 - B5)}{\%\ \text{of Sample}}$$

where-

D1 = ml of 0.025N Sodium Thiosulphate used for sample

D5 = ml of 0.025N Sodium Thiosulphate used for sample after incubation

B1 = ml of 0.025N Sodium Thiosulphate used for blank

B5 = ml of 0.025N Sodium Thiosulphate used for blank after incubation

2. Chemical oxygen demand (COD)

Principle: Chemical Oxygen Demand (COD is the oxygen required for the chemical oxidation of organic matter. The organic matter gets oxidized completely by Potassium Dichromate in the presence of Sulphuric Acid to produce Carbon dioxide and water. The Potassium dichromate consumed in the reaction gives the oxygen required for the oxidation of organic matter.

Requirements : Reflux Condenser, Flat bottom COD flask (500ml), Hot plate, Ferroin Indicator, Mercuric Sulphate, 0.025N Potassium dichromate (dissolve 12.259gm dried potassium dichromate in 1litre distilled water and add 120ml conc. sulphuric acid), Sulphuric Acid Reagent(add 10gm silver sulphate to 1000ml conc. sulphuric acid), 0.1N Ferrous ammonium sulphate (dissolve 39gm ferrous ammonium sulphat in 400ml-distilled water. Add 20ml conc. sulphuric acid and dilute to 1000ml)

Procedure

1. Use 0.4 gm Mercuric Sulphate in a COD flask and add 20ml sample.
2. Add glass beads followed by 10ml Potassium Dichromate Solution.
3. Add slowly 30ml Conc. Sulphuric acid (H_2SO_4) and mix well. If color turns green , add more Potassium Dichromate and Sulphuric acid.
4. Reflux for 2 hours, cool and wash down the condenser with distilled water.
5. Dilute for minimum 150ml, cool and titrate with Ferrous Ammonium Sulphate Solution (0.1N) using Ferroin Indicator till the color turns wine red / reddish brown.

6. Reflux the blank in the same manner using distilled water instead of sample.

Calculation :-

$$C.O.D\, mg/l = \frac{(A-B)*8000*N}{ml\ of\ sample}$$

Where, A = ml of Ferrous Ammonium Sulphate Solution (0.1N) for blank

B = ml of Ferrous Ammonium Sulphate Solution (0.1N) for sample

N = Normality of Ferrous Ammonium Sulphate Solution used.

❑❑❑

Chapter–7

Food Regulatory Acts and Laws

Over the past several years, there has been significant change in consumer perception on quality and food safety. Consumers are demanding healthy, nutritious and cost-effective products processed under controlled hygienic conditions. Food safety concern, in recent years, has become dominant and important. The requirement of food safety and quality assurance has never been so demanding as it is now. Globally food safety is being addressed independent of food quality. In anticipation of consumer demands on food safety, food sanitary standards at national and international levels are under constant review. Several inter governmental agencies / organizations are engaged in work related to quality assurance and food safety. Amongst these World Health Organization (WHO) and Codex Alimentarius Commission (CAC) play the major role in protecting public health and defining standards of quality and safety aspects of foods for regional and or world wide usage.

Regulatory Institutions

The quality environment in the Indian food processing sector revolves around compulsory legislation which specify minimum standard and certification system. Legal standards are generally made to have control over the quality of foods offered for sale and to safeguard the consumer from health hazards posed by practice of adulteration. There are basically three types of Indian standards namely

1. Food Safety and Standard act 2006

Food processing industry in India is presently governed by large number of food laws and/ or orders which are administrated by various ministries and are not considered condusive for effective fixation of food standard and their enforcement. The multiplicity of laws leads to greater confusion for investors, manufacturers, traders and consumers alike. Need for therefore felt for integration of such laws for effectively regulating the food quality. The food safety and standard act was passed by the parliament with the objective of ensuring safe food to the consumers. The act has 12 chapters, 101 clauses and 2 schedules. The major objectives of the act are :-

a) To provide single window system for matters related to standards and regulations.

b) Shifting emphasis from PFA to broader concept of food safety and management system.

c) Formulation of standards, samples, analysis, labelling under scientific advice.

d) Regulating manufacture, storage, distribution and sale to promote global trade.

e) Rationalize and strengthen existing enforcement mechanisms.

f) Providing graded civil and criminal penalties for food safety violations.

The Ministry of health and family welfare, Government of India has recently announced "Food Safety and Standard rule 2011" through Gazette of India Extraordinary Part II and III dated 19th January 2011. These rules are applicable and in force from 5th August 2011 all over India. The food safety and standard authority will be the supreme body to look into implementation and regulation of this act. The authority will consist of 7 members from different ministries (agriculture, commerce, health, food processing, small scale industries, legislative and consumers affairs); 2 from food industry; 3 food technologist/ scientists; 2 representatives from consumer organisation; 5 representatives from state; 2 representatives of farmers and 1 representatives from retailer organisation. The authority will have central advisory committee, scientific panels and scientific committee for providing scientific opinions. The state Govt. shall appoint commissioner of food safety for effective enforcement of requirements laid down in this act, rules and regulations made there under. The second schedule of the act repeals all other existing acts and orders except MMPO 1992. Central and state Governments. will establish appellate tribunal which will look into the matters related to violation of this act.

1. **Bureau of Indian Standards (BIS)** :- BIS is a statutory body established under the Ministry of Consumers Affair in 1986. It is the national standard organisation of India. It is responsible for laying down the Indian standards in consultation with the experts drawn from manufacturing units, research and technical institutions, purchase organisations and other concern parties. It deals in formulation of standards in processed food sector and implementation through voluntary and third party certification systems. These standards cover raw material permitted, their quality parameters, hygiene conditions of manufacturing, packaging and labelling requirements and are amended suitably from time to time. Manufacturers complying with these standards can obtain "ISI" or "BIS" certification and exhibit the same on their product package. In BIS standards, more emphasis is laid on the microbiological requirements. The standards are primarily focused on food processing establishments. Recently, the Bureau has launched a quality management system certification scheme to help strengthen the industry for their quality control programme. BIS operates following certification schemes :

   BIS product certification scheme.

   Quality scheme certification as per IS/ISO 9000 series.

Environmental management system as per IS/ISO 14000 series.

Hazard analysis and critical control point (HACCP) as per ISO-15000.

BIS has also started ECOMARK scheme for environmental friendly products. This was launched for protection of environment and ecology.

2. **AGMARK standard -** These standards are formulated by the Directorate of Marketing and Inspection, under Ministry of Agriculture, Govt. Of India. It enforces the Agriculture Produce (Grading and Marketing) act which was first enacted in 1937. Under this act, grade standards are prescribed for agricultural and allied commodities. These are known as "Agmark" standards. It categorises the commodities into various grades such as –special and standard. Grading under this act is voluntary and manufacturers who comply with the standards laid down in DMI are allowed to use "Agmark" logo on their products. "Agmark" gives the consumers an assurance of quality in accordance with the standards laid down. However, these standards are not mandatory. Presently, three dairy products namely: Ghee, Butter and dairy spreads are graded under this scheme.

Apart to these standards there are several other acts, standards and certification systems both for raw agricultural produce and processed foods that are either mandatory or voluntary from the core of compulsory legislation.

a) **Essential commodity act** – This act was enacted in 1955 and contains a number of control orders regulating the manufacturer, commerce and distribution of essential commodities including food.

b) **Standards of weight and measures act** – These rules formulated in 1977 specify certain obligatory conditions for all commodities that are in packed form with respect to declaration of quantities contained. The rules provide requirement for calibration, inspection, guidelines etc. for interstate trade. Under this act it is mandatory for the manufacturer to declare their name and address, month and date of manufacture, sale price etc. on the package. These rules are operated by the Directorate of Weight and Measures under the Ministry of Food and Civil Supplies, Govt. Of India.

c) **Consumer's protection act** – This act passed in 1986 provides protection to consumers against unfair trade practices of manufacturers and traders through right of consumers and applies to all goods and services. The act provides the consumers the right to safety, information, choose and education. An easy, speedy and inexpensive redressal is provided through complaint in district forum (up to Rs.1.00 lakh); State commission (1-10 lakh) and National commission (more than 10 lakh).

d) **Milk and Milk product order (MMPO)** – The Govt. Of India has setup an advisory board in 1992 for advice on production, sale, purchase and distribution of milk and milk products. In accordance with this order, all dairy plants processing more than 10,000 litres of milk per day or handling more than 500 tonnes of milk solids per annum are required to obtain registration. Dairy plants processing up to 75,000 litres per day or equivalent quantity of milk solids per annum need to be registered with the Directorate of Animal Husbandry in respective States, while those handling more than 75,000 litres per day have to register with Department of Animal Husbandry of Central Government. The registration is compulsory. This order has provision for definition of milk sheds for restriction on uneven procurement and marketing of milk, rules for production, hygienic conditions, packing, labelling, marketing, penalty etc. which will help in protection and growth of organised dairy industry.

e) **Essential commodities act** - A number of quality control orders have been issued under Essential Commodities Act such as Milk & Milk Products Order (MMPO), Meat Product Order (MPO) and Vegetable Oils Control Order (VOCO). These orders are mandatory and are primarily meant for regulating the hygienic conditions. The Government of India's latest amendment of MMPO (March'2002) from the Department of Animal Husbandry and Dairying (Ministry of Agriculture), in its fifth schedule now includes Sanitary/Hygiene measures for Quality and Safety of food which shall be applicable even for the business handling <10,000 l of milk per day. Together these orders are included in a single order named the Food Products Order (FPO).

f) **Export (Quality control and Inspection) act** – This act works under the Department of Commerce and operated by the Export inspection Council of India was enacted in 1963. The Export inspection Council of India (EICI) is responsible for operation of this act. It envisages compulsory inspection of the manufacturers premises, human hygiene, quality of raw material purchased, hygienic manufacturing practices followed, quality assurance programmes followed, packaging and labelling etc. at the production centre for the exportable product. Export inspection agencies help EICI in arranging inspection of food manufacturing plants to verify implementation of approved guidelines. The dairy establishments, under the act, need to maintain records on parameters related to hygiene and food safety and make it available to EIC regularly. A certificate to the effect has to be obtained before the consignment is cleared for export for each batch. The act covers compulsory pre-shipment examination of large number of exportable commodities. The quality control and inspection of export product is administered through net work of officials located at the main production centers and port of shipment. EICI is also implementing residue Monitoring Plan (RMP) for the plants that are approved for exporting food products.

g) **Environmental protection act (EPA) :** Authorities in India have established to deal with the provisions of EPA in relation to GMOs and GM foods. There is a Genetic Engineering Approval Committee (GEAC) which functions under the Department of Environment, Forestry and wild life for approval of large scale use of micro-organisms and recombinants for research and industrial production from the angle of environmental safety. The committee is responsible for approval of proposals related to release of genetically engineered organisms and products.

h) **Food safety/sanitary requirements:** legal aspectsWith the expanding significance of global trade after the establishment of WTO, a free trade in safe food across international borders was agreed. WTO paved the way for several multilateral agreements on trade, which include Technical barriers to trade (TBT) relating to all goods and Sanitary and Phytosanitary agreement (SPS) concerning agricultural products. The TBT agreement covers technical standards of food like product description and labelling while SPS agreement covers food aspects related to health, like food additives, pesticide residues, veterinary drugs and hygiene practices. The quality aspects of all primary products of plant and animal origin fall under SPS measures. In regard to food, the term Sanitary and Phytosanitary measure means any gauge applied to protect human and animal life or health with in the territory of the member countries from risk arising from food additives, contaminants, toxins or disease causing organisms in foods, beverages, or feed stuffs. For animal life and health, the SPS agreement recognizes the standards adopted and recommended by the International Office of Epizooites (OIE) and for plant life and health, those recommendations of the International Plant Protection Convention (IPPC). The organizations Codex, OIE and IPPC are known as three sisters. Only those measures that are necessary to send measurement must be based on scientific principles. The SPS agreement specifically refer to standards, guidelines and recommendation of the CAC as the bench mark for dispute settlement as well as and basis for harmonization of national food legislation. This was done to avoid the trade restriction arising due to Variation in the procedures of national food control systems involving monitoring and sampling, detection and analytical methods, certifications, import/export regulation, application of standards and food safety regulations. There fore, while carrying out the process of harmonization, it is imperative that national legislations on food safety are based on sound science.

Hygiene code for effective sanitation during production and processing of milk and milk products.

Codex Committee on Food Hygiene (CCFH) is elaborating the code on General Principles of Food Hygiene. The draft code of "Hygiene practices for Milk and Milk Products" was documented at the 35th session of the CCFH held from 27th Jan'2002 to 1st Feb'2003 at Orlando, Florida (USA). It is currently at step 5 of the procedure for its final approval by CAC. This code is a complete document on hygienic conditions required during the entire food chain. The application of these codes will improve the sanitary quality of dairy food, raw materials/ingredients in terms of prevention from adulteration of food contact surfaces, prevention of cross contamination, control of process operation, control of employee health and exclusion of pest etc. The codex is flexible enough to be applicable to different milk production systems including smallholder dairying system prevalent in our country.

ISO 9000 Certification for Dairy and Food enterprises

The WTO agreement has made the World smaller for trading. In the borderless world trade it has become imperative to signify the quality capability of a company by quality systems. In this direction it has become very essential to obtain necessary international recognition to exhibit our commitment to quality which can be achieved by obtaining ISO certification. It is a network of national standard institute of 163 countries with a central secretariat in Geneva, Switzerland that coordinates the system. It is the Worlds largest developer of international standards. It is a non-governmental organization that acts as a bridge between public and private sector. ISO standards are based on consensus among the experts in the field. The policy development committee provides strategic guidance for development of standards and provide reference frame work to concern shareholders by agreeing on specification and criteria to be applied consistently in classification of materials, manufacture and supply of product, testing & analysis and provision of services. This facilitates trade and transfer of technology. It provides technical base for health, safety and environmental legislation. These standards are voluntary and may become market requirement as in case of ISO 9000 quality management system. They are globally relevant and are useful everywhere in the World. Majority of ISO standards are highly specific to products, materials or processes. ISO standards provide guidance on good management practices. These management systems help the organization to manage its processes/ activities in order that its product/services meet the set goals like consumer quality requirements, meeting regulations and environment objectives. It provides with international state-of-art model to follow in a systemized manner for more efficient and effective application. The ISO series does not specify quality levels, rejection percentage, waste percentage etc. However, its implementation definitely, leads to continuous improvement of these parameters and quality levels in a systematic manner in dairy/food enterprise. ISO started its work on quality standardization through its technical multidisciplinary international Committee. The Committee formulated standards of quality management system popularly known as ISO 9000 series.

ISO 9000 : "Guidelines for selection and use of standards on Quality Systems" – This provides guidelines for selection and use of the series of International Standards on Quality systems that can be used for internal and external quality management purpose.

ISO 9001 : "Model for quality assurance in design, development, production, installation and servicing" – This standard is applicable in contractual situations when - The contract specifically requires design effort and the product requirements and stated in performance terms or they need to be established.

ISO 9002 : "Model for quality assurances in production and installation". This Standard is applicable in Contractual situations when - The specified requirements for product are stated in terms of an established design or specification.

ISO 9003 : "Model for quality assurance in final inspection and test". This International Standard specifies quality system requirements for use where a contract between two parties requires demonstration of a supplier's capability to detect and control the disposition of any product non-conformity during final inspection and test.

ISO 9004 : "Quality management and quality system elements – Guidelines". This International Standard describes a basic set of elements to provide guidance for developing and implementing a sound quality management system. The selection of appropriate elements contained in this standard.

ISO 14000 - After a success of ISO 9000 series, the international organization for standards developed a comprehensive set of standards for environmental management. This is a series of international standards on environment management system (EMS). It was developed as a result of Rio submit on environment held in 1992. The series is designed to cover the whole area of environmental issues in the global market place. These standards are made to encourage a cleaner, safer, healthier World for us all. ISO 14001 is a generic standard which represent the core set of standards for designing and implementing an effective environment management system. It helps the organization to minimize their operational negativity. The major objective is to promote more effective and efficient environmental management and to provide useful and usable tools that are cost effective, system based, flexible and reflects the best organizational practices available for gathering, interpreting and communicating environmentally relevant information. It offers a source of guiding and adapting EMS based on the best universal practices. ISO 14001 can be employed to measure and document their environmental impact. Both ISO 9000 and ISO 14000 series can be implemented side by side. EMS that meets ISO 14001 requirements can be externally audited and certified by an accredidated certification agency.

ISO 22000 – It is a food safety management standard used to establish and certify food safety management system (FSMS). ISO 22000 defines a set of food safety management system requirements. In order to establish food safety management system (FSMS) the organization needs to establish food safety policy by developing a food safety policy document; procedures and records required by this standard. One has to make sure that top managers are committed and communicate their support for food safety policy and ensure that the food safety policy is implemented throughout the organization. The managers plan the development ; documentation ; implementation ; maintenance ; evaluation and modification of your FSMS. The responsibilities and authorities are defined and communicate to all personnel responsible for reporting FSMS problems and only designated personnel are given the responsibility and authority to solve FSMS problems. To start with a team leader is chosen and authorized to establish ; manage , implement, report and maintain effective internal and external communication arrangements to address food safety issues.

Hazard Analysis Critical Control Point (HACCP) – The sole objective of HACCP concept is to produce safest possible food items for consumers, considering each and every possible hazard well in advance. Incorporation of the Hazard Analysis Critical Control Point (HACCP) system helps in identifying the potential hazards in the process operations based on viability/proliferation of the contaminants and their control measures. It determines the critical control points, defines critical limits and their monitoring and thus guards the food against any food safety hazards. It must be developed by the individual organization tailored to its individual product, processing, packaging, storage, distribution and marketing. The use of HACCP was given on international dimension by CAC on food hygiene as an annexure to the Codex "Recommended International Code of Practice: General Principles of Food Hygiene". Revised guidelines for the application of HACCP System (ALINORM 03/13A, Appendix II) have been endorsed by CAC to make HACCP more flexible for less developed business.

The food safety programme is designed to ensure that the food product is safe through controls at all stages of procurement, processing, manufacture and shipment. HACCP has been recognized internationally as bench mark and is an ideal tool to provide assurance for food safety.

The development and implementation of HACCP based food safety management programme involves a multi disciplinary team efforts in gathering experience and relevant expertise in different areas of food production, processing, public health, environment health, toxicology etc. The food manufacturing unit must identify and describe raw material ingredients, processing parameters, preservation techniques, packaging materials used and their impact on food safety.

Hazard analysis – It is the process of collecting and evaluating information on hazard and conditions leading to their presence. It encompasses physical, chemical and microbiological agents in food processing. The objective is to list all hazards expected to occur at each step in the food chain. It involves identification of potential hazard, significance, sensitivity and probability of occurence, establish preventive measures and result tabulation.

Process flow – The team develops a detailed process flow diagram indicating critical steps of control. It is used as a basis of hazard analysis and must contain sufficient technical details. The hazard and conditions leading to their presence are significant to food safety and therefore, should be addressed in food safety management system.

Critical control points and limits (CCP) - After establishing significance of hazards, CCP are determined. It is the procedure where a food safety hazard can be prevented or reduced to acceptable limits. The critical limit is the maximum / minimum value to which the hazard must be controlled to prevent the occurrence of the identified food safety hazards. The critical limit should be specified and preventive measures defined. The criterion used could be measurement of temperature, time, moisture, pH and sensory parameters.

Monitoring – It is important and must detect loss of control and be able to provide timely warning so that corrective action could be taken before the limit is exceeded. For effective monitoring a trained personal may be identified and made responsible and be able to understand purpose and importance of monitoring.

Varification / Validation - To provide assurance on day-to-day basis, the system must include verification procedure. Auditing methods like random sampling and analysis may be employed to determine if the HACCP system is working effectively and correctly. Validation is an essential feature for successful operation of food safety management system. It is defined as "scientific and technical process for determining if the CCP and associate critical limits are adequate to control likely hazards". Validation is performed at different stages during initial implementation and subsequent operations of the system to ensure indented functioning.

Implementation – After the HACCP plan is designed, verified and validated it is implemented as stand alone or exercised through ISO-9001 as integrated system. The main steps involved in implementation are – awareness training for the entire team, placement of documented system at work place, determining Implementation time frame, internal audit and appraisal system and obtaining certification from accredited agencies.

HACCP system is an internationally accepted proactive and preventive mechanism for food safety. It adds reliability to quality and safety of dairy products. Most of the dairies in India have adopted it as an integrated system.

Globalization of Indian food safety laws

With the expanding significance of global food trade, the need for a more universal food code will be paramount in the near future. The Food and Agriculture Organization (FAO) and the World Health Organization (WHO) have set up a Codex Alimentarius Commission (CAC) and subsidiary bodies that present a unique opportunity for all countries to join the international community in formulating food standards and working towards their global implementation.

India's contribution in world food trade is relatively insignificant despite high foodstuff production of 500 million tons annually. One of the major factors for such low exports has been the food quality and food safety aspect. To ensure a strong presence in global markets, India needs to meet the challenges and keep pace with the development at international level market. The international food standards/guidelines/codes related to food hygiene and sanitation formulated by the codex are accepted by the WTO, as the reference points for the global food trade. In due course of time, most of the national food standards, if not all, are likely to be harmonized with those of codex and the dairy industry would be required to comply with those standards. Development of the dairy portion of the codex is in process and will be of significance in governing the hygienic processing practices and recommendations relating to compliance with dairy product standards that can be adopted globally.

There are various bodies dealing with standards formulation activities. There are a number of food safety regulations (mandatory or voluntary) that are implemented by various Ministries/ Departments within India. The review of multiple laws is necessary to have uniform and logical approach for regulating the quality of food and for harmonizing with the international regulations. Various ministries are taking the following actions.

Harmonization of BIS standards - BIS is the largest body for formulating standards for various food items. It is voluntary. The Indian standards IS:15000:1998 on Food Hygiene – HACCP system and guidelines for its application is equivalent to FAO/WHO, CAC document. The Ministry of Civil Supplies and Consumer Affairs has brought out a paper for consideration of Committee of Secretaries (COS). The paper recommends that BIS should formulate standards for all food items in the country. This will be a major step towards harmonization of food laws and is still under consideration of COS for finalization.

Harmonization of PFA standards- The primary national public agency for India is Prevention of Food Adulteration Act 1954 and its amendments from time to time. The Central Committee for Food Safety Standards (CCFSS) of the Director General of Health services, Ministry of Health and Family Welfare are responsible for its implementation. The task force constituted by the Prime Minister has submitted its report which is under the consideration of the government. The task force has advocated promotion of food safety and quality and has further made the following suggestions:

- Food Regulation Authority (FRA) to be set up to formulate and update food standards for domestic and export market.
- FRA should replace the PFA to conform to international standards.

❑❑❑

Suggested Reading

Arora, K.L. and Bhatia, K.L. (1990). Formal titration method for determining protein content in ice-cream. Indian Journal of Animal Sciences, 60 (10) 1252-1255.

Arora.K.L. (1995) Determination of milk solids in flavoured milk using BIS lactometer. Indian J Dairy & Biosciences, 6 : 7-10.

Bector,B.S., MotiRam and Singhal,O.P. (1998). Rapid platform test for detection/adulteration of added urea in milk. Indian Dairyman, 50(1), 59-62.

Bishop,J.R., Senyk,G.F. and Duncun,S.E. (1993). Detection of antibiotic/drug residues inmilk and dairy products. In : Standard methods for examination of dairy products. 16th edition, American Public Health Assoc., New York.

Brich,G.G. and Mwangelwa,O.M. (1974). Colorimetric determination of sucrose in sweetened condensed milk products. J Sci. Food and Agric., 25(11) : 1355-1362.

Codex Alimentarious Committee, (1997). Recommended international code of practices. General principles of food hygiene. CAC/RCP -1 and 22.

Fagel,A. and Ambrus,A. (2000). Recent development in field of validation methods. IDF Bulletin No. 351, pp 4-10.

IDF Standard (1991). Milk and milk products. Recommended method for determination of organochlorine compounds (pesticides). IDF 75C pp 12 -15.

IS: 1224 Part 1- Milk Part II, Milk Products (1977). Indian Standards. Determination of fat in milk by the Gerber Method IS: (First Revision). Bureau of Indian Standards, Manak Bhawan, New Delhi.

IS: 1479 (Part-II) (1981). Indian Standards. Methods of tests for Dairy Industry. Chemical Analysis of Milk. Bureau of Indian Standards, Manak Bhawan, New Delhi.

IS: 11964. (1987). Indian Standards. Method for detection of free acidity in casein. BIS, Manak Bhawan, New Delhi.

IS (1992). Indian Standards. Specification for milk powder and skim milk powder.

IS: 14433 (Part 1). (1997) Indian Standards. Specification for infant milk substitutes. BIS, Manak Bhawan, New Delhi.

IS: 1656 (1997). Indian Standards. Specification for milk-cereal based weaning foods. (Third Revision). BIS, Manak Bhawan, New Delhi.

IS: 2488 (Part1-Part5), (1998). Methods of Sampling and Test for Industrial Effluents. BIS, Manak Bhawan, New Delhi.

IS: 2785 (1999). Indian Standards. Specification for natural cheese (Hard variety), Process cheese, Process cheese spread and soft cheese. Bureau of Indian Standards, Manak Bhawan, New Delhi.

IS: 5887 (Part 3), (1999). Indian Standards. Methods for detection of bacteria responsible for food poisoning, General guideline on methods for detection of *Salmonella* (Second Revision). BIS, Manak Bhawan, New Delhi.

IS: 5887 (Part 7), (1999). Indian Standards. Methods for detection of bacteria responsible for food poisoning, General guideline on methods for detection for detection of *Shigella*. BIS, Manak Bhawan, New Delhi.

IS: 5401 (Part 1,) (2002). Indian Standards. Microbiology – General guidelines for the enumeration of coliform, (First Revision). BIS, Manak Bhawan, New Delhi.

IS: 14988 (Part 2), (2002). Indian Standards. Microbiology of food and animal feed stuffs – Horizontal method for detection and enumeration of *Listeria monocytogenes*, BIS, Manak Bhawan, New Delhi.

IS: 5887 (Part 8/Sec.1), (2002). Indian Standards. Methods for detection of bacteria responsible for food poisoning. Horizontal method for enumeration of coagulase positive Staphylococci (Staphylococcus aureus and other species). BIS, Manak Bhawan, New Delhi.

IS 13334 (Part 1), (1992). Standard for skim milk powder (special grade). Indian standard Institution, manak Bhawan, new Delhi.

IS :3508 (1966). Indian standards methods of sampling and tests for ghee. BIS, manak Bhawan, new Delhi.

IS: 11202 (1984). Indian standards. Method for determination of lactic acid and lactate contest in milk powder and similar product, BIS, Manak Bhawan, New Delhi.

ISI Handbook of Food Analysis Part 1 – General Methods IS:18 (Part 1) 1980.

ISI Handbook of Food Analysis Part IX – Dairy Products IS: 18(Part IX) 1981.

ISO (1975) Determination of lactic acid and Lactate content. ISO:3497(E).

Joel F.Ross (1996) Total quality management. Vicinity Book International, New Delhi.

Majonnier, T. and Troy, H.C. (1925). Technical Control of Dairy Products, 2nd edn. p 108.

Majonnier Bros. & Co., Chicago, CF Laboratory Manual, 1959, 3rd edn. p 283, MIF Washington, D.C.

Mannual for Dairy Chemistry (1982) ICAR -N.D.R.I., Publication.

Mannual for Dairy Microbiology (1972) ICAR - N.D.R.I., Publication.

Offical methods of analysis of the AOAC., (1984) 14th edition.,published by AOAC., Virginia,USA.

Official Methods of Analysis of AOAC International – 16th Edition – Vol. 1 and 2. Pub.- AOAC International, 2200 Wilson Blvd., Varginia, USA.

Prevention of Food Adultration Act - as revised in 2006

Rao,G.S. and Arora,K.L. (1996) Colorimetric determination of sucrose in ice-cream. Indian J dairy sci., 49(4) : 293-298.

Richmond, H.D. (1941). Richmonds Dairy Chemistry, J. G. Davis and F.J. MacDonald, 5th edn. p 160, Charles Griffen & Co., London.

Singh,R.S., Chander,H. and Kaul,A. (1990) Detection of antibiotics in milk. Pub. No. 258, NDRI,Karnal.

Sohrab, M. , (1997) – HACCP through ISO-9000 quality management system. Indian Food Industry, 16:3.

Sohrab, M., (1999) – Risk assessment as pre-requsite for implementation of HACCP. Indian Food Packer, July-August. Standard Methods for Examination of Dairy Products. Ed. Robert T. Marshall, 16th Edition , 1993. Pub.- APHA, Washington D.C., USA.

Sohrab, M. , (2000) – Adoption of Food safety management system- A challenge for dairy industry. Bev.Food World, 27:7.

Standard Methods for examination of Water and Waste Water (1995), 19th Ed. APHA, Washington D.C.

The Environmental (Protection) Act.,1986 (2000).Universal Law Publishing Co.Pvt.Ltd.,Delhi-110033

Wadhawa,B,K.; Bhattacararjee,S.K.; Bindal,M.P.; Chabra,A. and Mann,B. (2000) Standardization of package of practicees for analysis of organochloro pesticide residues in milk. Indian J dairy sci., 53(3) : 169-176.

❑❑❑

Appendix A

Standards of fat & SNF for raw milk in India. (FSSAI rules 2011)

States	Buffalo milk		Cow milk	
	Fat	SNF	Fat	SNF
Andhra Pradesh	5.0	9. 0	3.5	8.5
Arunachal Pradesh	5.0	9.0	3.5	8.5
Assam	6.0	9.0	3.5	8.5
Bihar	6.0	9.0	3.5	8.5
Delhi	6.0	9.0	3.5	8.5
Goa	5.0	9.0	3.5	8.5
Gujarat	6.0	9.0	3.5	8.5
Haryana	6.0	9.0	4.0	8.5
Himachal Pradesh	5.0	9.0	3.5	8.5
Jammu & Kashmir	5.0	9.0	3.5	8.5
Karnataka	5.0	9.0	3.5	8.5
Kerala	5.0	9.0	3.5	8.5
Madhya Pradesh	5.0	9.0	3.5	8.5
Maharashtra	6.0	9.0	3.5	8.5
Manipur	5.0	9.0	3.5	8.5
Meghalaya	6.0	9.0	3.5	8.5
Mizoram	5.0	9.0	3.0	8.5

Contd.

Nagaland	5.0	9.0	3.5	8.5
Orissa	5.0	9.0	3.0	8.5
Punjab	6.0	9.0	4.0	8.5
Rajasthan	5.0	9.0	3.5	8.5
Sikkim	6.0	9.0	3.5	8.5
Tamilnadu	5.0	9.0	3.5	8.5
Tripura	5.0	9.0	3.5	8.5
Uttar Pradesh	6.0	9.0	3.5	8.5
West Bengal	6.0	9.0	3.5	8.5
Union Territories				
Andaman & Nicobar	6.0	9.0	3.5	8.5
Chandigarh	6.0	9.0	4.0	8.5
Dadar & Nagarhaveli	5.0	9.0	3.5	8.5
Daman & Diu	5.0	9.0	3.5	8.5
Lakshadweep	5.0	9.0	3.5	8.5
Pondicherry	5.0	9.0	3.5	8.5

Definitions of different milk types and related processes as per FSSAI - 2011

A. Types of Milk

1. **Milk** - Milk is normal mammary secretion derived from complete milking of healthy animal without either addition thereto or extraction there from. It shall be free from colostrums. Milk of different classes / designation shall confirm to the standards laid down.

2. **Standardised milk** – Standardised milk means cow milk or buffalo milk or sheep or goat milk or a combination of these milk that has been standardised to requisite fat and SNF percentage by adjustment of milk solids. It shall be pasteurised and shall show negative phosphatase test.

3. **Toned milk** – Toned milk means the product prepared by admixture of cow or buffalo milk or both with fresh skim milk; or by admixture of cow or buffalo milk or both, that has been standardised to requisite fat and SNF

percentage, by adjustment of milk solids. It shall be pasteurised and shall show negative phosphatase test. When fat or dry non-fat-milk solids are used, it shall be ensured that the product remains homogenous and no deposition of solids take place on standing.

4. **Double Toned milk** - Double toned milk means the product prepared by admixture of cow or buffalo milk or both with fresh skim milk; or by admixture of cow or buffalo milk or both, that has been standardised to requisite fat and SNF percentage, by adjustment of milk solids. It shall be pasteurised and shall show negative phosphatase test. When fat or dry non-fat-milk solids are used, it shall be ensured that the product remains homogenous and no deposition of solids take place on standing.

5. **Full Cream Milk** – Mean milk or a combination of buffalo or cow milk or a product prepared by combination of both that has been standardised to requisite fat and SNF percentage, by adjustment / addition of milk solids. Full cream milk shall be pasteurised. It shall show negative phosphatase test. It shall be packed in clean, sound and sanitary containers properly sealed so as to prevent contamination.

6. **Skimmed Milk** – Means the product prepared from milk from which almost all the milk fat has been removed mechanical.

7. **Mixed Milk** - Means the product of combination of milk of cow, buffalo, sheep, goat or any other milch animal and may be combination of any of this milk, which has been made and conforms to the standards.

8. **Flavoured Milk** – By what ever name called, may contain nuts (whole, fragmented or ground), chocolate, coffee or any other edible flavour, edible food colours, and cane sugar. Flavoured milk shall be pasteurised, sterilised or boiled. The type of milk shall be mentioned on the label.

9. **Recombined Milk** – Means the homogenised product

prepared from milk fat, non-fat milk solids and water. Recombined milk shall be pasteurised and shall show negative phosphatase test.

10. **Boiled Milk** – Means milk that has been brought to boil.

11. **Cream** – Including sterilised cream means the product of cow or buffalo milk or combination thereof. It shall be free from starch and other ingredients foreign to milk. It may be of following three categories, namely

Low fat cream – containing milk fat not less than 25% by weight.

Medium fat cream- containing milk fat not less than 40% by weight

High fat cream- containing milk fat not less than 60% by weight.

12. **Malai** – means the product rich in butterfat prepared by boiling and cooling cow or buffalo milk or a combination therefore. It shall contain not less than 25% milk fat.

B. Related Processes

1. **Pasteurization** – The term pasteurization / pasteurized and similar terms shall be taken to refer to process of heating every particle of milk to at least 60° C and holding at such temperature for at least 30 minutes, or heating to at least 71.5° C and holding at such temperature for at least 15 seconds or an approved temperature-time combination hat will serve to give a negative phosphatase test. All pasteurized milk shall be cooled immediately to a temperature of 10° C or less.

2. **Sterilization** – The term sterilization when used in association with milk, means heating milk in sealed containers continuously to a temperature of either 115° C for 15 minutes or at least 130° C for a period of one second or more in a continuous flow and then packed under aseptic conditions in hermetically sealed containers to ensure preservation at room temperature for a period not less than 15 days from the date of manufacture.

Standard for different classes of milk (All India standard)

Class of milk	% Fat (minimum)	% SNF (minimum)
Full cream milk	6.0	9.0
Standardized milk	4.5	8.5
Recombined milk	3.0	8.5
Toned milk	3.0	8.5
Double toned milk	1.5	8.5
Skimmed milk	Not more than 0.5	8.7
Mixed milk	4.5	8.5

Note: Milk without any designation of class shall confirm to standard for Buffalo milk.

BIS requirement for different types of milk

Milk type	Fat (%)	SNF (%)	TS (%)	Cream index	LA (%)	Neutralizer	Sucrose (%)
Skim milk	1.5 max	8.7	10.5	-	0.14	Absent	-
Double Toned Milk	1.5	9.0	10.5	-	0.14	Absent	-
Toned milk	3.0	8.5	12.0	-	0.15	Absent	-
Standard milk	4.5	8.5	13.0	-	0.15	Absent	-
Full cream milk	6.0	9.0	15.0	-	0.16	absent	-
Sterilized milk	3.5	8.5	14.0	20	0.18	Absent	-
Flavoured milk	3.0	8.5	12.0	20	0.18	Absent	6.0
Condensed milk (plane)	NLT 9.0	-	31.0	-	0.35	-	-
(Sweetened)	9.0	-	31.0	-	0.35	-	40.0
(Skimmed)	NMT 0.5	-	26.0	-	0.35	-	-
Partly skimmed	>3.0 & <9.0	-	28.0	-	0.35	-	-
UHT condensed	NLT 8.0	-	26.0	-	0.35	-	18.0
UHT cond. skimmed	NMT 0.5	-	20	-	0.35	-	18.0

Note: - NLT = Not less than, NMT = Not more than

Standards for milk products (FSSAI- 2011)

1. Fat rich products

Product type	Moisture (%)	Fat (%)	Curd (%)	Salt (%)
Malai	-	25	-	-
Cream (low fat)	-	25	-	-
Medium fat	-	40	-	-
High fat	-	60	-	-
Table butter	16	80	1.5	3.0
Cooking butter	-	76	-	-
Table spread	16-56	41-80	-	2.5

2. Concentrated dairy products

Product type	Fat (%)	Protein (%)	Milk solids (%)	Fat on dry matter (%)	Moisture (%)
Khao	-	-	-	30.0	-
Channa	-	-	-	50.0	70.0
Paneer	20 - 25	-	-	50.0	70.0
Low fat paneer	15	-	-	17.0	70.0
Evaporated milk	8	34	26	-	-
Partially skimmed	1-8	34	20	-	-
skimmed	1	34	20	-	-
High fat	15	34	27	-	-
Sweetened condensed milk	9	34	31	-	-
Partially skimmed	3-9	34	28	-	-
skimmed	1	34	26	-	-
High fat	16	34	30	-	-

3. Fermented milk products

Product type	Milk TS / s.n.f (%)	Fat (%)	Protein (%)	Ash (%)	Sugar (%)
Chakka	TS 30.0	33	30	3.5	-
Full fat	20.0	38	30	3.5	-
skimmed	28.0	5	60	5	-
Srikand	58	8.5	9	0.9	72.5
Full fat	58	10	7	0.9	72.5
Yoghurt	s.n.f. 8.5	8	3.2	-	-
skimmed	8.5	0.5	3.2	-	-
sweetened	8.5	3	3.2	-	6
fruit	8.5	1.5	2.6	-	6
Partially skimmed	8.5	0.5 – 3.0	3.2		

4. Frozen dairy products

Product type	Fat (%)	Protein (%)	Milk solids (%)	Wt./Vol (g/l)
Ice cream	10	3.5	36	525
Low fat	2.5	3.0	26	475
Medium fat	2.5 - 10	3.5	30	475
Milk lolly	2.0	3.5	20	-

5. Powder dairy products

Product type	Moisture (%)	Fat (%)	Protein (%)	Carbo-hydrate (%)	Ash (%)	Insolubility index (ml)
WMP	4.0	26.0	34.0	-	7.3	2.0
SMP	5.0	1.5	34.0	-	8.2	2.0
Partially skimmed	5.0	1.5 - 26	34.0	-	8.2	2.0
Infant milk food	4.5	18	12.0	-	8.5	2.0
Whey powder	5.0	2.0	10.0	61.0	9.5	-
Cream powder	5.0	42.0	34.0	-	-	-
Cereal based complementry food	5.0	7.5	15.0	55.0	5.0	-
Infant formula	4.5	18	10 -16	-	8.5	2.0
Followup formula	4.5	18-27	13.5	-	8.5	2.0

6. Cheeses

Cheese type	Fat-on dry-matter (%)	Moisture (%)
Extra hard	32.0	36.0
Hard pressed	48.0	39.0
Semi hard	40.0	45.0
Soft	20.0	80.0
Mozzerella	35.0	60.0
Pizza	35.0	54.0
Processed	40.0	47.0
Processed cheese spread	40.0	60.0

7. Butter oil and Anhydrous milk fat

Parameters	Butter oil	AMF
BR reading (40° C)	40-44	40-44
Moisture % (NMT)	0.4	0.1
Milk fat % (NLT)	99.6	99.8
RM Value (NLT)	24.0	24.0
FFA % (Oleic acid) NMT	0.4	0.3
Peroxide value (mEq/kg fat) NMT	0.6	0.3
Boudounin test	Negative	Negative

8. Ghee

Sl. No.	State / Union Territory	B R reading at 40°C	R M Value (Min.)	Percentage of	
				FFA as oleic acid (max.)	Moisture (Max.)
1	Andhra Pradesh	40.0 to 43.0	24	3	0.5
2	Andaman & Nicobar Islands	40.0 to 43.0	24	3	0.5
3	Arunachal Pradesh	41.0 to 44.0	26	3	0.5
4	Assam	40.0 to 43.0	26	3	0.5
5	Bihar	40.0 to 43.0	28	3	0.5
6	Chandigarh	40.0 to 43.0	28	3	0.5
7	Chattisgarh	40.0 to 44.0	26	3	0.5
8	Dadra and Nagar haveli	40.0 to 43.0	24	3	0.5
9	Delhi	40.0 to 43.0	28	3	0.5

contd.

10 Goa	40.0 to 43.0	26	3	0.5
11 Daman & Diu	40.0 to 43.5	24	3	0.5
12 Gujarat				
12a Areas other than cotton tract areas	40.0 to 43.5	24	3	0.5
12b Cotton tract areas	41.5 to 45.0	21	3	0.5
13 Haryana				
13a Areas other than cotton tract areas	40.0 to 43.0	28	3	0.5
13b Cotton tract areas	40.0 to 43.0	26	3	0.5
14 Himachal Pradesh	40.0 to 43.0	26	3	0.5
15 Jammu & Kashmir	40.0 to 43.0	26	3	0.5
16 Jharkhand	40.0 to 43.0	28	3	0.5
17 Karnataka			3	0.5
17a Areas other than Belgaum district	40.0 to 43.0	24	3	0.5
17b Belgaum district	40.0 to 44.0	26	3	0.5
18 Kerala	40.0 to 43.0	26	3	0.5
19 Lakshwadeep	40.0 to 43.0	26	3	0.5
20 Madhya Pradesh				
20a Areas other than cotton tract areas	40.0 to 43.0	26	3	0.5
20b Cotton tract areas	41.5 to 45.0	21	3	0.5
21 Maharashtra				
21a Areas other than cotton tract areas	40.0 to 43.0	26	3	0.5
21b Cotton tract areas	41.5 to 45.0	21	3	0.5
22 Manipur	40.0 to 43.0	26	3	0.5
23 Meghalya	40.0 to 43.0	26	3	0.5
24 Mizoram	40.0 to 43.0	26	3	0.5
25 Nagaland	40.0 to 43.0	26	3	0.5
26 Orisssa	40.0 to 43.0	26	3	0.5
27 Pondicherry	40.0 to 44.0	26	3	0.5
28 Punjab	40.0 to 43.0	28	3	0.5
29 Rajasthan				
29a Areas other than Jodhpur District	40.0 to 43.0	26	3	0.5
29b Jodhpur District	41.5 to 45.0	21	3	0.5
30 Tamil Nadu	41.0 to 44.0	24	3	0.5
31 Tripura	40.0 to 43.0	26	3	0.5
32 Uttar Pradesh	40.0 to 43.0	26	3	0.5
33 Uttarakhand	40.0 to 43.0	26	3	0.5

contd.

34	West Bengal				
34a	Areas other than Bishnupur sub division	40.0 to 43.0	28	3	0.5
34b	Bishnupur sub division	41.5 to 45.0	21	3	0.5
35	Sikkim	40.0 to 43.0	28	3	0.5

9. Casein and Casienate

Parameters	Rennet Casien	Acid Casien	Casienate
Moisture %	12.0	12.0	8.0
Fat %	2.0	2.0	2.0
Protein %	84.0	90.0	88.0
Lactose	1.0	1.0	1.0

Microbiological standards

a. Microbial counts

Products	Total count	Coliform	Yeast & Moulds
Pasteurized milk, cream, flavoured milk	30,000 – 50,000/g	<10/g	-
Sterilized/UHT milk, cream, flavoured milk, evaporated milk	-	-	-
Sweetend condensed milk	500-1500/g	<10/g	<10/g
Table butter	10,000-50,000/g	<50/g	<50/g
Dreid powders-milk, cream, whey, edible casein	40,000-50,000/g	<50/g	-
Khoa	50,000-1,00,000/g	<90/g	<100/g
Channa & Paneer	3-5 lakh/g	<90/g	<250/g
Dahi, chakka, srikhand, yoghurt	-	<50/g	<100/g
All Cheeses	-	<500/g	<100/g
Processed cheese, Cheese spread	50,000-75,000/g	<10/g	<10/g
Ice-cream & Frozen dessert	2-2.5 lakhs/g	<100/g	<10/g

b. Pathogenic bacterial counts

Product	E coli	Salmonella	Staphylococus aureus	Clostridium preferinges	Listeria monocytogen
Pasteurized milk, cream, flavoured milk	Absent/g	Absent/25g	<10/g	-	Absent/g
Sterilized/UHT milk, cream, flavoured milk, evaporated milk	-	-	-	-	Absent/g
Sweetend condensed milk	Absent/g	Absent/25g	<10/g	<100/g	Absent/g
Table butter	Absent/g	Absent/25g	<50/g	-	Absent/g
Dreid powders-milk, cream, whey, edible casein	Absent/g	Absent/25g	<10/g	<100/g	Absent/g
Khoa	<10/g	Absent/25g	<100/g	-	Absent/g
Channa & Paneer	<10/g	Absent/g	<100/g	-	Absent/g
Dahi, chakka, srikhand, yoghurt	Absent/g	Absent/25g	<100/g	-	Absent/g
All Cheeses	<10/g	Absent/g	<1000/g	<10/g	Absent/g
Processed cheese, Cheese spread	Absent/g	Absent/25g	<10/g	<100/g	Absent/g
Ice-cream & Frozen dessert	Absent/g	Absent/25g	<10/g	-	Absent/g

Note : See sampling guidelines for microbiological test.

Appendix B

Maximum Residual Limits of Toxic Metals

Toxic metals	**Max.level (ppm)**
Arsenic	Milk-0.1; Products-1.1
Cardimum	1.5
Copper	30
Lead	2.5
Mercury	1.0
Methyl-mercury	0.25
Tin	250
Zinc	**50**

Adapted from Varsheny,2004

Maximum Residue Limit (MRL) or tolerances for veterinary drugs

Drug	Codex (MRL)	MRL and safe/Tolerance levels ug/kg milk (i.e. ppb)			
		Canada (MRL)	European Union (MRL)	UK Vety. products Committee (MRL)	FDA (Safe/ Tolerance)
Penicillins:					
Amoxycillin			4		10/10
Ampicillin		10	4		10/10
Benzylpenicillin			4	4	5/10
Cloxacillin		30	30		10/10
Dicloxacillin			30		
Oxacillin			30		
Penicillin Penethamate		6	4		5/0
Cephalo -sporins :					
Ceftiofur Cefquinone Cephapirin	100		100 20		50*1000@ 20/20
Chlorte-tracycline Oxytetracycline Tetracycline	100	150	100 100 100	20 100 100	30/0 30/0 80/0
Macrolides:					
(Neo) spiramycin			200		
Erythromycin		50	40		50/0
Tylosin			50		50/50
Aminogly -cosides:					
Dihydrostrep-tomycin	200	125	200		125/0
Gentamycin	100	125	100		30/0
Neomycin (+framycetin)	500	250	500		150/150
Spectinomycin	200		200		30/0
Streptomycin		125	125,200		125/0
Quinolones: Marbofloxancin			75		

Contd.

Polymyxins: Colistin			50		
Ansamycins+ naphthalene			60		
Aromatics:					
Chloramphenicol			0	10	0/0
Novobiocin		125	-		100/100
Sulphonamides:					
Total Group as such Dapsone			100 0	100	
Sulphadiazine					10/0
Sulphadimidine	25				10/0
Sulphadoxine Sulphamerazine		10	100		-10/0
Sulphamethoxine		10	100		10/0
Sulphathiazole					10/0
Sulphamethazine			10	100	10/0
Baquiloprim			30		
Trimethoprim			50		

Adapted from -(Honkanen-Buzalski and Reybroeck, 1997, Mitchell *et al.,* 1998).

Maximum residual limits (MRL) for some pesticides (IDF,1997)

Pesticide	Codex (ppm)	PFA (ppm)
2,4-D	0.01	0.05
Alderin & dieldrin	0.006	0.15
Carbendazim	0.05	0.10
Carbofuran	0.05	0.05
Chlodane	0.002	0.05
Chlopyrifos	0.02	0.01
Cypermthrin	0.05	0.01
DDT	0.02	1.25
Diphenylamine	0.0004	–
Dimetholate	0.05	–
Dithiocarbamate	0.05	–
Endosulphan	0.004	–
Ethion	–	0.5
Fenitrothion	0.002	0.05
Fenvalerate	0.10	0.01
Heptachlor	0.006	0.15
BetaHexacholrocylohexane	0.01	0.01
DeltaHexacholrocylohexane	–	0.02
Methidathion	0.001	–
Monocrotophors	–	0.02
Phenthoate	–	0.01
Phorate	0.05	0.05
Primiphos-methyl	0.05	0.05
Profenophos	0.01	–
Tricholofon	–	0.05
Thiabendazole	0.2	–
Trizophos	**0.01**	–

❑❑❑

Appendix C

List of Approved Analytical Laboratories Used for the Purpose of Rml

1. Delhi Test House, G.T.Karnal Road, Azadpur, Tele fax: 011-27437327 / 27456917 Delhi – 110033. Mail: deltest@bol.net.in
2. Shriram Institute for Industrial Research, 19,University Road, Delhi – 110007.
3. Shriram Institute for Industrial Research Plot No. 14 – 15, Sadarmangala Industrial Area Whitfield Road, Bangalore – 560048. Ph: 080-28410165 Mail: srisrb@vsnl.com
4. Arbro Pharmaceuticals Ltd. 4/9 Kirti Nagar Industrial Area New Delhi – 110015
5. SGS India Pvt. Ltd. 1/509 A, Old Mahabalipuram Road Ph: 044-24962822 Thoraipakkam, Chennai – 600096 Fax: 044- 24963075
6. Choksi Laboratories Limited 6/3, Manoramganj, Indore – 452001 (MP) Ph: 0731-493592/3

❑❑❑

Colour Plates

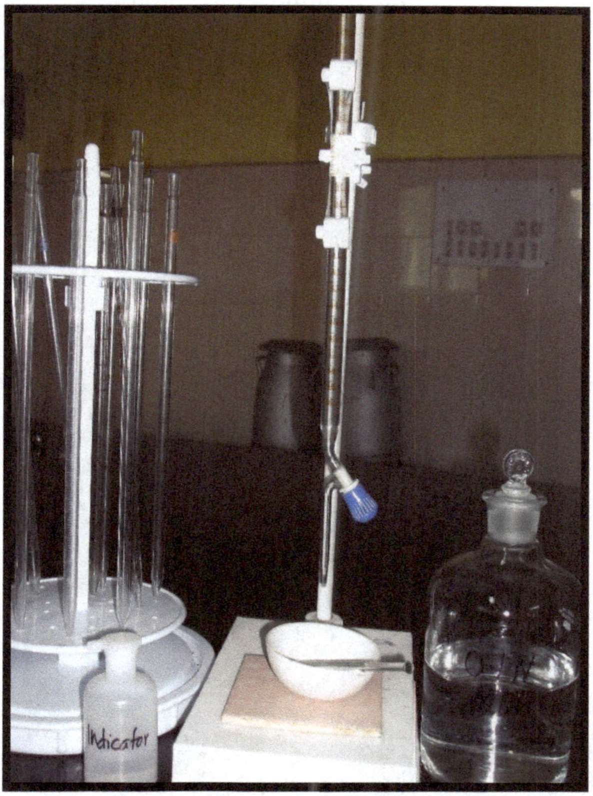

Acidity testing set up

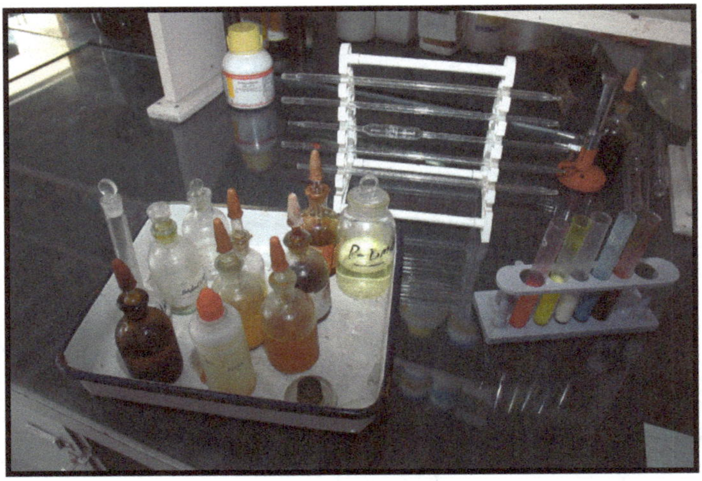

Testing of milk adulteration

Lactometer test

Gerber fat testing set

Sampling Utensils

Laminar flow cabinet

Chemical lab

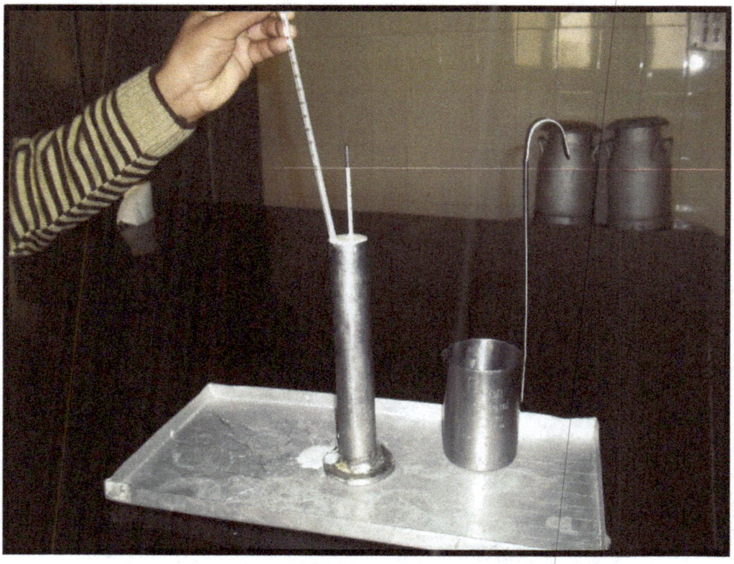

SNF testing

Gerber Centrifuse

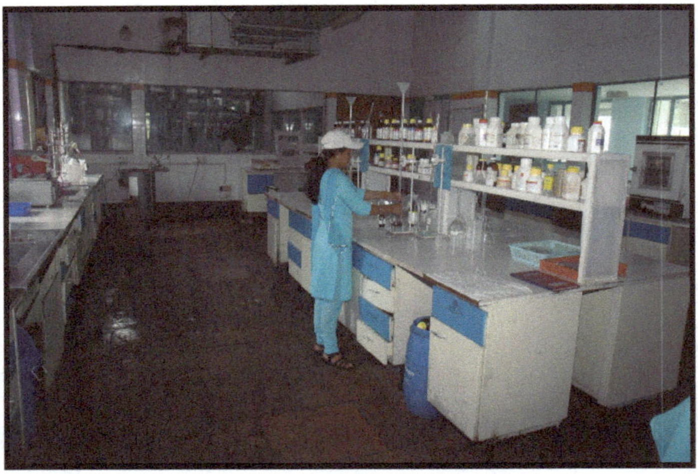

View of Quality assurance lab

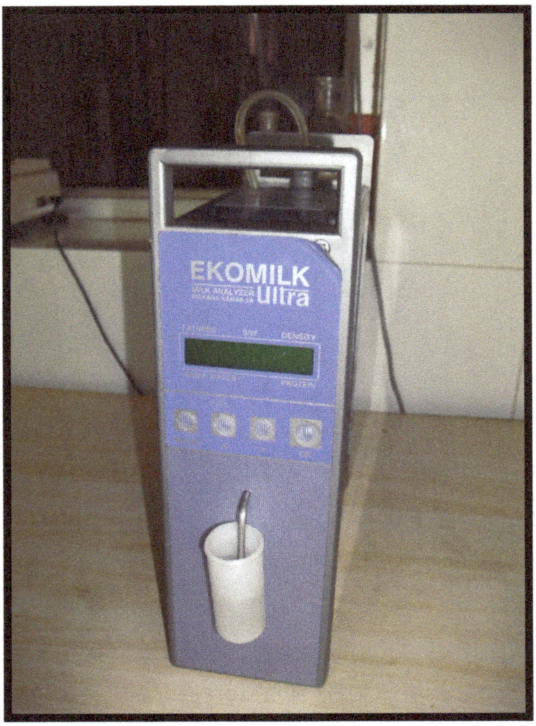

Electronic fat tester

Microbiological lab

Milk Reception dock lab

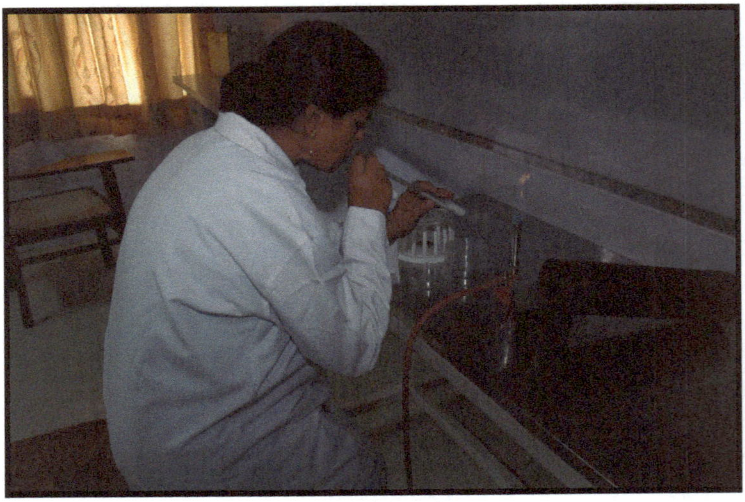

Microbial plating

❑❑❑

Index

N

O

P

Q

R

S

T

U

V

W

Y

MIX

Papier | Fördert gute Waldnutzung

FSC® C083411

Zeitfracht Medien GmbH
Ferdinand-Jühlke-Straße 7
99095 Erfurt, Deutschland
produktsicherheit@kolibri360.de